THE GARBH CHRIOCHAN A' DEAS LOST PLACE-NAMES SURVEY

VOLUME 2

THE LOST PLACE-NAMES OF

ARDNAMURCHAN

AND

MOIDART

COMPILED BY

J E KIRBY

WITH TEARLACH MACFARLANE, ALISTAIR MACINTYRE, CATRIONA HUNTER, JOAN MADDEN

and

MARILYN GASCOIGNE

2015

High Cross, Eilean Fhianain.

ISBN: 978-1-910205-69-3

Additional names, stories, corrections, other relevant information and orders may be sent to
dahlkirby@gmail.com

Front cover: The march dyke between Faskadale and Achateny, with *Port Bàn* (left) and *Guraban* (right), looking
across to Rum and Eigg.

Back cover. Glen Moidart looking NE across *Loch nan Lochan* to *An Eas Briadha* 'the splendid waterfall' in Glen
Forslan.

CONTENTS

THE GARBH CHRIOCHAN A' DEAS

DISTRICTS WITHIN THE SOUTHERN ROUGH BOUNDS

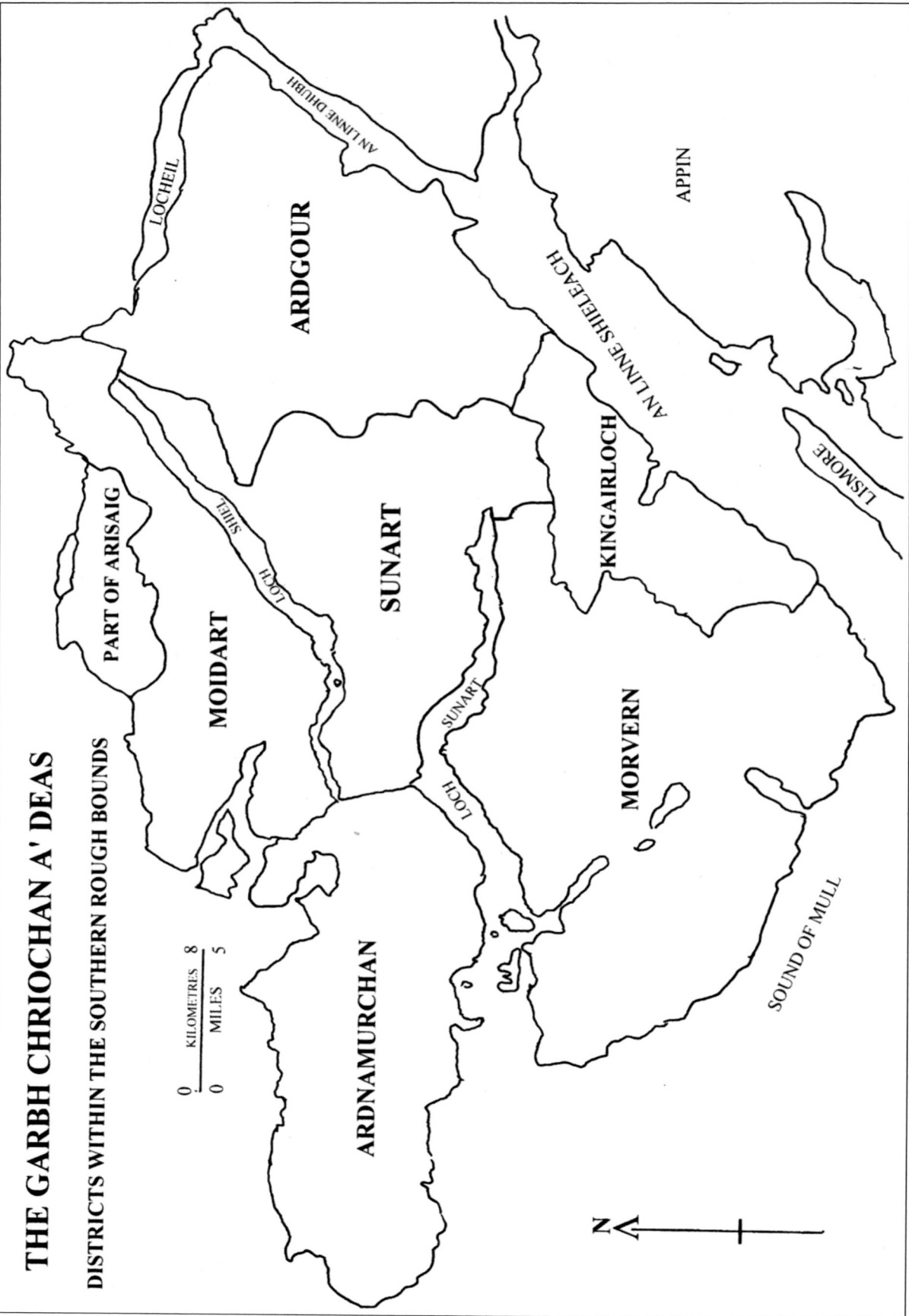

KILOMETRES 8
0
MILES 5
0

LOCHEIL

AN LINNE DHUBH

ARDGOUR

APPIN

AN LINNE SHILEACH

LISMORE

PART OF ARISAIG

LOCH SHIEL

MOIDART

SUNART

KINGAIRLOCH

ARDNAMURCHAN

LOCH SUNART

MORVERN

SOUND OF MULL

N

4

GLOSSARY

Bailie	The laird's agent or steward.
Blackwood	A charcoal burner's term for wood which produces second rate charcoal i.e. birch, willow, rowan, aspen.
Bowman	A man holding a farm on a Steelbow.
Bronze Age	A sub-division of prehistory, characterised by the introduction of smelting copper and the production of metal weapons and jewellery from copper alloys, commonly called 'bronze'. C. 2,500-1,000 BC.
Brother-german	Full-blood brothers; having the same parentage.
Burn	The Scots word for Gaelic *allt* 'stream, brook' or Cumbrian 'beck', but here used as a loan word into English.
Byre	Cow shed.
Cashel	From Old Irish *caisel* a 'stone built rampart round a monastery'.
Changehouse	Inn.
Chilick; chlitech	A ½ farthing land.
Chirurgeon	Surgeon or doctor.
Coals	Charcoal.
Cottar	A sub-tenant given a house and a small area of land in exchange for labour service. Usually allowed to keep a cow or two, plus followers.
Creel	A wickerwork basket or describing something incorporating wickerwork.
Croft	A small farm held directly from the landowner but usually too small to sustain a family.
Developed shieling	A shieling on good agricultural land which was later enclosed, cultivated and occupied on a permanent basis.
Dùn	A fort, usually small enough to have been roofed, and of Iron Age date.
Excambion	A legal exchange of land.
Factor	A person appointed to act on someone else's behalf; an agent.
Fank; sheepfank	Sheepfold. An enclosure made from stone, turf or stake and rice, used for working with and sorting out sheep.
Feal	Turf; feal or fale dyke, one made of turf.
Feu	A piece of land on which the buyer owns the buildings but the land remains the property of the landowner (superior), for which an annual rent or feu duty is paid.
Glebe	Land granted to a clergyman as part of his benefice.
Gralloch	Animal entrails or the act of removing them.
Ground Officer	The Bailie.
Haugh	Field or pasture at the side of a loch, river or burn.
Infeftment	Granting of legal possession.
Iron Age	A sub-division of prehistory characterised by the introduction of smelting and manufacture of iron tools, the use of the rotary quern and people speaking Celtic languages. Spans the interface between the later prehistoric and early historic periods i.e. the 'Roman Iron Age/Early Christian Period'. C. 700 BC-550 AD.
Kelp	Ashes or fused slag produced by the incineration of seaweed in a kelp kiln and valued for its alkali or iodine content (but not properly used for the seaweed itself).
Kiln	A structure for drying corn (corn kiln) or for the manufacture of kelp (kelp kiln) or lime (lime kiln).
Lazybeds	G. *feannag*. Hand dug cultivation ridges which have a flat surface on which to grow potatoes or oats, with the furrows providing drainage. Dwelly describes the term as 'a southern odium on the system of farming in Gaeldom, where soil was scarce and where bog-land could not be cultivated in any other way'.
Loch	Gaelic for 'lake' but here adopted as a loan word into English.
Lochan	A small loch, adopted here as a loan-word into English.
March	An estate or property boundary. A March dyke, marking the boundary between one township and another.

Mart	Market. Auction Mart, where livestock was sold.
Meeting House	After the Reformation the Presbyterian Church converted the former Roman Catholic and Episcopalian churches into Sermon or Meeting Houses by removing the alter from the east end and replacing it with a Communion Table in front of the north or south wall.
Mesolithic	A sub-division of prehistory; the 'Middle Stone Age', from c. 7,500-4,000 BC, 'Scotland's First Settlers', although there would almost certainly have been a small population of Upper Palaeolithic 'Old Stone Age' people in Scotland during the inter-glacial periods and as the Ice Age passed.*
Miserable hutt	A creel house or building.
Mr.	The use of the title Mister, G. *Maighstir* denotes a man, usually a minister of the church, with a university degree
Nein	From Gaelic *nighean* 'daughter of'.
Neolithic	A sub-division of prehistory; the 'New Stone Age', the first farmers in Britain c. 4000-2500 BC, introducing grain crops processed using saddle querns, pottery and complex chambered burial tombs.
Offices	Farm buildings, usually a barn, byre and corn kiln.
Park	Field.
Penticle	Part of a township that could be let separately; a smallholding let to a sub-tenant.
Persoun	Parson, Minister.
Pitstead	Charcoal burner's platform.
Port	A place where stones have been cleared from the beach to allow boats to land safely. A 'boatslip'.
Residenter	A resident, particularly one of long standing.
Rotary quern	A stone hand mill for grinding corn.
Relict	Widow.
Run-rig	A system of holding a share of a township's rigs in scattered and intermingled blocks, which were subject to periodic, and sometimes annual redistribution between the tenants.
Sermon House	See Meeting House.
Shielings	Summer pasture to which livestock was taken during the summer months to exploit the scattered grazing land and keep the stock away from the arable ground. Often used here to denote the summer residencies and associated huts, pens and store houses used by the people that tended the animals on the hill.
Skerry	A rock in the sea, generally visible at most stages of the tide.
Smiddy.	Blacksmith's shop or forge.
Stake and rice	A fence made of stakes interwoven with rods, poles or brushwood.
Steading	Animal and storeage buildings around a farm.
Steelbow	A form of tenancy in which a landlord provides the tenant with stock, grain, implements etc., under contract that the equivalent should be returned at the end of the lease.
Stell	A small circular sheepfank without internal divisions, so only used as a shelter.
Tack	A lease or tenancy, especially the leasehold tenure of a farm, mill etc.
Tacksman	A tenant farmer who holds a lease or tack of a township, and who is able to sub-let some of the land for other people to work. A prominent member of the clan and the backbone of the chief's fighting force
Town; Township	A farm, with cultivated inby land and hill pasture.
Transhumance	The seasonal migration of livestock and their attendants to suitable grazing grounds (the shielings) away from the arable land.
Vaccary	Cow-town. A ranch.
Vallum	From Latin *vallum* or *vallum monasterrii,* an earthen bank and outer quarry ditch surrounding or enclosing a monastery and constituting a spiritual and legal boundary between the monastic establishment and the secular world outside.
Violer	A Fiddler, often in the service of the clan chief.

Vitrified fort	A timber-laced fort, usually Iron Age in date, which when set on fire, produced such a high temperature as to cause the stones to melt and fuse.
Wadset	A lease granted to a wealthy tacksman in exchange for a cash loan, which could be redeemed when the laird raised the necessary finance.
Waste	Land lying untenanted and uncultivated.
Wedder	Wether; a Castrated tup (ram).
Wintertown	Permanantly occupied houses and steadings where the people that attended to the crops lived and to which the people of the shielings returned during the winter months. G. *Baile-geamhraidh* 'infield, the low and cultivated fields of a township', but literally 'the winter town'.
Yell cows, yell sheep	Barren animals. Yell beast, a barren cow.

*At the time of going to press, news has come in of the discovery of upper Palaeolithic occupation of the Ahrensburgian culture on *Rubha Port an t-Seilich* on the east coast of Islay which takes the the first settlers in Scotland back a further 3,000 years to c.13,000 BC. This is likely to be the most significant archaeological discovery in Scotland this century.

Cruciform stone with tapered shaft and side-arms c. 0.50m high by 0.22m across the arms.
At the centre of the east face there is an incised Latin cross, 28mm high and 22mm in span.
The old parish church of Island Finnan.

ACKNOWLEDGEMENTS

I would like to thank the staff at the Heritage Lottery Fund for the intial funding for the project, and for much forbearance in renewing the contract agreement over a number of years. The delays were entirely my own fault. Thanks are due to members of the Sunart Community Council who, over the years gave me their support in backing the project from the very beginning. Cathlin Macaulay and the staff of the School of Scottish Studies, University of Edinburgh gave me access to their sound archives and associated maps prepared by Ian Fraser and Bill Nicolaison in the 1960s and 70s, and I am most grateful for their pioneering work in this field of research. The staff at both the National Archives of Scotland and the National Library of Scotland have been most friendly and helpful in sourcing manuscripts and maps over many years, while the staff at both the Ardnamurchan Library, Strontian and the Fort William Library sourced many obscure works of reference, and allowed me to borrow books from the Local Collection.

The Sunart Oakwoods Research Group gave me free access to their archives and survey work, and provided supplementary match-funding. The Lochaber Forest District of Forestry Commission Scotland very kindly provided maps of the area. The National Museums Scotland allowed me to use their images of the drop from the crozier found on Dalelia, while the copper alloy ring-brooch from the Green Isle is here illustrated for the first time. Dr. Sandra Evans very kindly provided images of several distant townships which I no longer had the energy to visit myself. Marilyn Gascoigne was an excellent (and uncomplaining) companion on many expeditions, and some of our archaeological discoveries will continue to appear in the pages of *Discovery and Excavation in Scotland* for a number of years to come. Ken Bowker very kindly established the Group's web-site.

Professor Richard A. V. Cox of Sabhal Mòr Ostaig acted as referee for many of the place-names, while Tearlach MacFarlane offered additional information through his knowledge of the local dialect, places, place-names, people and events. Gordon Barr also provided a lot of information on the Moidart place-names and I made full use of the excellent collection of names on the *Comann Eachdraidh Mùideart* web-site. Alistair MacIntyre and Joan Madden also added a considerable amount of knowledge concerning the place-names of the Acharacle and West Ardnamurchan areas, but I must accept responsibility for the inevitable mistakes and more fanciful interpretations.

The School of Scottish Studies, University of Edinburgh gave me permission to quote from their sound archives, as did the families of the recorded tradition-bearers; Morag MacDonald for Alasdair Cameron 'North Argyll', Mary Khan for Hugh MacKenzie, and the families of John Cameron, Duncan Cameron and Archie Campbell, whom I failed to trace. Other tradition-bearers, informants and assistants included Malcolm MacMillan, Joey MacKenzie and Alistair MacPhail, who are sadly no longer with us, Duncan Cameron, Mary Cameron, Alan MacNaughton, John Dye, Alasdair Carmichael, Angus Peter Maclean, Jeff Carr, Hugh and Bridget Cameron, Peter Madden, Gill Calver, Owain Kirby, Magnus Kirby and Alasdair MacLean of Keystore, Strontian.

Finally I would like to thank Richard Burkitt and all the team at "For The Right Reasons" for their guidance, skill and forebearance in steering both volumes through the printing and publication phases of production,

THE LOST PLACE-NAMES OF

THE ACHARACLE AREA

SET IN A HISTORICAL, ARCHAEOLOGICAL

AND CULTURAL CONTEXT

ARDNAMURCHAN TOWNSHIPS WITHIN THE ACHARACLE COMMUNITY COUNCIL AREA

CONTENTS

Shepherd's marker on Arivegaig

Title Page of the Acharacle Community Council area within Ardnamurchan:
Eas Bun na h-Aibhne and the estuary of the River Shiel, with Casteal Tioram and Moidart beyond.

ARDNAMURCHAN TOWNSHIPS

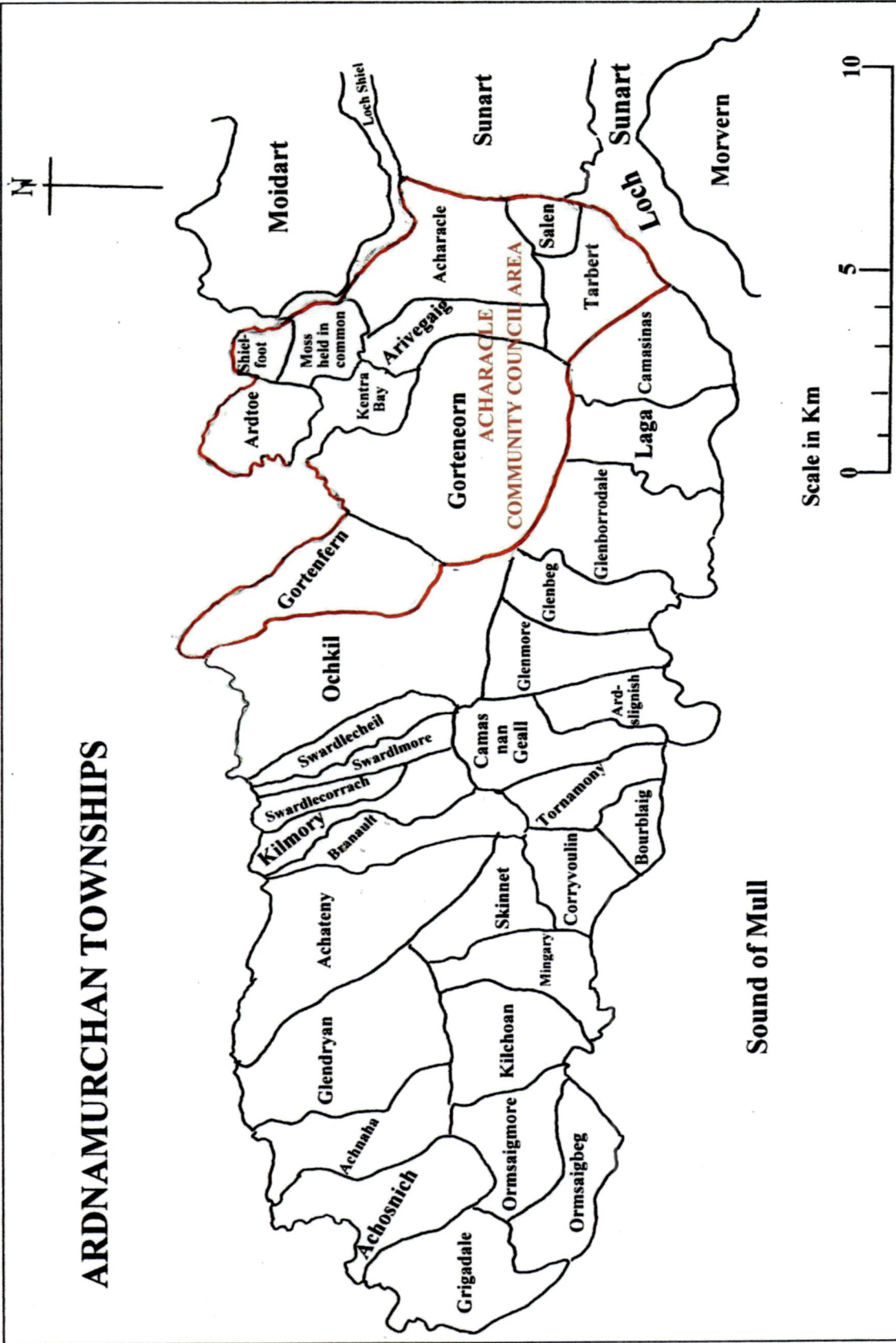

N

Moidart

Loch Shiel

Sunart

Acharacle

Shiel-foot

Moss held in common

Arivegaig

Salen

Ardtoe

Kentra Bay

Gorteneorn

ACHARACLE COMMUNITY COUNCIL AREA

Tarbert

Loch Sunart

Morvern

Gortenfern

Laga

Camasinas

Ochkil

Glenborrodale

Glenbeg

Swardlecheil

Swardlmore

Glenmore

Swardlecorrach

Camas nan Geall

Ard-slignish

Kilmory

Branault

Tornamony

Achateny

Skinnet

Corryvoulin

Bourblaig

Glendryan

Kilchoan

Mingary

Achnaha

Achosnich

Ormsaigmore

Ormsaigbeg

Grigadale

Sound of Mull

Scale in Km

0   5   10

11

ACHARACLE : Àth Tharacheal : Tarracail's ford

1541, Aherkill (47); 1610, Aherkill (35); 1651, Aherkill (11); 1667, Aucherkill (5); 1667, Airkill (5); 1694, Aharakill (6); 1694, Aharickle (6); 1716, Aharkile (37); 1722, Aharkil (27); 1723, Aharkill (10); 1723, Acharakle (7); 1737, Acharackile (13); 1742, Aharakeil (12); 1828, Acharacle (24); 1875, Acharacle (8); 2002, Acharacle (9).

Acharacle from *Rubha an t-Sabhail Ghuail*

Valuation: Assessed as a 2½ merk land in 1610, and a 5 penny land by 1723.

Area: Of the 1716.93 acres there were 30.47 arable, 67.07 pasture, 409.99 moss and 1206.29 moor in 1807. In 1828, Thom. Anderson wrote: *The crofts have been newly arranged & a lease given. Enclosures & temporary Houses commenced* (24). In the 1890s the Medical Officer of Health for Argyll reported that most of the houses in Acharacle were 'mud huts'.

Tenants: 1541, Alester McAngus McKane (47); 1691, John McPherson and his wife Mary NcIlveile, with their children, Margaret, Mary, Katherine and Donald 6); John McIlveile with his wife Mary NcOlvernie and daughter Mary (6); 1716, Ewn Cameron and his son Alexander, Angus Cameron, Allan McAlester VcDhoile, John McIllespick dhui, John Cameron Crofts and John McInnish VcGuen (37); 1737, Let with Ardtoe to John Cameron (13); 1743, John mc Conchie vic Dhuilt and Donald mc Dhonichie vic Dhuil his brother, alias mc olonichs, Dugald mc Dhonichie Roy mc oleny, Duncan and Dougald mc olonys his sons, Duncan mc Allester vain mc olony, John mc Ian vic Dhuil mc olony (12); 1828, 8 tenants leasing farms on a year to year basis and 7 crofters with leases from 1828-1842 (24); 1847, 6 tenants and 8 crofters (40). 1864, 7 farmers and 1 crofter were 'cleared' (43).

Acharacle from Coire Cruinn

Population: 1723, 9 families, with 10 men, 14 women,19 children, total 43.

Settlements: NM 6786 6778, Acharacle.

Shielings: NM 6882 6578. *'Lochan na Dubh-Leitir'.*

Acharacle (detail). National Records of Scotland, RHP72/1-8, Plan of Ardnamurchan and Sunart, Argyll, 1806.

Placenames:

1. **Acharacle (9). NM 6786 6778.** G. *Àth Tharracail* 'Tarracail's ford' or 'Thorkild's ford'.

2. **Ru na Nell (1). NM 6928 6820.** G. *Rubha nan Eala* 'the headland of the swan̲s'. Named *Rubha na h-Eala* 'the headland of the swan' on OS maps.

3. **Rudha a' Chorrain Duibh (8). NM 6789 6832.** G. *Rubha a' Chorrain Duibh* 'the headland of the black sheep'.

4. **Innis an Druim (2). NM 6862 6645.** G. *Innis an Droma* 'the pasture or enclosure of the ridge'.

5. **Gobsheiloch (3). NM ?** G. *Gob Seilich* 'the point of the willow'.

6. **Hope's Road (4). Rathad Mòr Hope (33). NM 6840 6765.** This name was used for the Salen to Acharacle road but it's use now appears to be confined to the short section leading down from the Loch Shiel Hotel to the pier.

7. **Glac Eachainn (30). NM 6717 6815.** G. *Glac Eachainn* 'Hector's pass'.

8. **Raineach Mary Hamish (30). NM 6840 6684.** An old croft to the east of the Acharacle fank.

9. **Rie Vannie (30). NM 688 664.** G. *Ruighe a' Bhainne* 'the shieling (or hill pasture) of the milk'.

10. **Gorten Martin (30). NM 6764 6717.** G. *Goirtean Mhàrtainn* 'Martin's field'.

11. **Glac Eilean (30). NM 6729 6737 ?** G. *Glac Ailein* 'Allan's gully'.

12. **Cnoc Alluicht (30). NM 6740 6757 ?** Perhaps G. *An Cnoc Falaichte* 'the hidden hill', - here *Fhalaichte* from dative use.

13. **Larach an Taigh mor (30).** Location not known. G. *Làrach an Taighe Mhòir* 'the ruins of the big house'.

14. **Glac a' Muic (30). NM 6891 6826 ?** G. *Glac na Muice* 'the hollow of the pig'.

15. **Clach Carnanais (30).** Location not known. ? *Clach Carnanais,* 'the stone of the ? cairn', or G. ?*Clach nan Càrnanaich* 'the stone of the Highlanders'.

16. **Ard Beithe (33). NM 6914 6690.** G. *Àird Bheithe* or *Àird na Beithe* 'the promontory of the birch'.

17. **Lochan a Cura (1). NM 6735 6843.** G. *Lochan na Cura* 'the lochan of the sheep'. Named *Lochan a' Churraidh* on OS maps.

A turf house (foreground) and field system on Acharacle, looking over Arivegag to Gorteneorn

Acharacle Observer Post

18. **Acharacle Observer Corps Post (31). NM 6879 6599.** A brick-built Observer Post code-named JON 3 which was manned around the clock by two teams during WW II, with at least 5 men in each team. It was connected by telephone to other posts at Strontian (JON 1) and Kingairloch (JON 2), and with the Control Centre in Oban. During the whole war, only one German aircraft was seen, a Heinkel bomber, presumably on its way to bomb the aluminium factory in Fort William.

19. **Lochan na h-Inghinn (46). NM 6750 6565.** G. *Lochan na h-Ingheann* 'the lochan of the young woman'. Named *Lochan na Dubh-Leitir* 'the lochan of the black slope' on OS maps.

20. **Loch Aithneachd (46).** G. *Loch Aithneachd* from *Aithneachd* ?'knowledge, humanity, recognition, discernment'. An alternative name for *Lochan na h-Ingheann* above.

Acharacle from Coire Cruinn

ARDTOE : Àird Tobha: the height or point of the headland

1541, Ardto (47); 1610, Ardto (35); 1667, Ardto (5); 1690, Ardtoo (6); 1723, Ardtoe (7); 1723, Artoe (10); 1784, Ardtoe with the Isle of Riska (39); 1807, Ardtoe (1); 1828, Ardtoe (24); 1875, Ardtoe (8); 2002, Ardtoe (9).

Valuation: Assessed as a 2 merk land in 1541, 1610 and 1667, and along with 'Waterfoot' as a 4 penny land in 1723.

Area: In 1807 the 1306.855 acres consisted of 10.245 arable, 9.460 cultivated with the spade, 754.080 moor & pasture and 533.070 of moss, which was held in common with Shielfoot.
Alex. Lowe wrote:
 '*This is a very rough piece of pasture, and being small is better suited for black cattle than sheep. The arable land is very good being chiefly composed of calcareous Earth or shell sand, considerable beds of which are exposed in a Bay here. This may be used as a Manure round the skirts of the moss with good effect. This ought to be in one possession. The Moss here added to the measure is common , and is very bad, being mostly flow moss. Here the sand has been prevented from blowing over the bit arable land by Stake and Rice*' (wickerwork fences made from small roundwood).

Ardtoe

Tenants: 1541, lying waste (47); 1690, Dugald Cameron with his wife Mary and children John, Alexander, Duncan and Voir (6); 1716, Ewn bane McVarrish, Duncan Cameron, Donald bane McInnish VcDhoil ruay, Ewn McKendrig, Donald Ean oig, John Cameron, his son, Donald Johnstoun and John McKuarig (37); 1737, John Cameron (13); 1739, Alex'r McDonald, Schoolmaster (15); 1742, John and Duncan mc olonys (12); 1807, Capt. Don. Cameron and one other; 1828, 8 tenants on a year to year basis (24); 1847, 8 tenants, 8 crofters and 5 more in Newton (40). 1852, 8 tenants were evicted for non-payment of rent (44, 45).

Population: 1723, 4 families, with 8 men, 11 women, 9 children, total 28.

Settlements: NM 6286 7020. Ardtoe.
  NM 6490 7004. Gobshealach.
  NM 6455 7062. Newton of Ardtoe,

Shielings: Not known. Ardtoe was probably too small to require shielings.

Placenames:

1. **Ardtoe (9). NM 6286 7020.** G. *Àird Tobha. Àird* 'height or point' *and tobhá* 'headland'.

2. **Faoil Vannan (1). NM 6532 6880.** G. *An Fhadhail Bàn* 'the white sea ford'.

16

Ardtoe

National Records of Scotland, RHP72/1-8, Plan of Ardnamurchan and Sunart, Argyll, 1806.

3. **Ru na Haorrie (1). NM 6458 6894.** G. *Rubha na h-Aoraid.*? The name of a headland, perhaps with *aoraidh* 'worship'.

4. **Illan an Dou (1). NM 6485 7100.** G. *Na h-Eileanan Dubha* 'the little black island'. Named *Eilean an Eididh* on OS maps.

5. **Port na Hinnse (1). NM 6389 6984.** G. *Port na h-Innse* 'the port of the little meadow'.

6. **Port na Glac (1). NM 632 702.** G. *Port na Glaice* 'the port of the narrow glen'.

7. **Ru Linnen (1). NM 6240 7064.** G. *Rubha* ? A headland. Named *Rubha Luinngeanach* on OS maps.

8. **Ru Fasan a feorag (1). NM 6220 7154.** G. *Rubha Fàsadh nam Feòrag* 'the headland of the squirrel's stance'? Named *Rudha Fassadh nam Feòcullan* on OS maps. *Feòcullen* 'polecat'. This is a very unlikely habitat for a squirrel, and a polecat is unlikely.

9. **Port Canna (1). NM 6282 7080.** G. *Port ...?* The name of a 'port' but the specific is difficult to read.

Sàilean Dubh

10. **Camus Linga (1). NM 6270 7075.** G. *Camas na Luinge* 'the bay of the ship'. In July 1732, Duncan Cameron in Ardtoe bought *'the Berge or vesshell the [blank]of Uist from Lauchlan Machk Vaistir Alistir vic Innes oig alias McDonald in South Uist then owner thereof... for the agreed price of Thirty Pound stirlin money'.* Lauchlan was allowed to take the boat, laden with 12 tonnes of oak timber provided by Duncan, to Ireland, where he was to sell the timber as payment for the vessel. Lauchlan sold the timber as agreed, and the boat a second time, for which he was fined £40 Stirling by the Vice Admiral Court of Argyll (NRS AC20/2/5(2). This bay was probably a safe anchorage where relatively large trading vessels like this could be kept.

11. **Ru Linga (1). NM6262 7084.** G. *Rubha na Luinge* 'the headland of the ship'.

12. **Poll Dubh (31). NM 6476 6900.** G. *Am Poll Dubh* 'the black pool' or 'black inlet'.

13. **Port Huin Bain(?) (1) NM 6273 7074.** G. *Port an Uain Bhàin* 'the port of the grey lamb'.

14. **Skeir Adhie (1). NM 6185 7125.** Named *Sgeir an Eididh* on OS maps and this may just be a phonetic rendering of 'Skeir Adhie' 'the skerry of the clothing'. Possibly G. *Sgeir an t-Séididh* 'the windy skerry'

15. **Shea Rauch or 'Sixed Oar'd Rock' (1). NM 6211 7159.** G. *Sia Ràmhach* 'a six-oared boat'. Does the name preserve the memory of a six-oared boat which came to grief on this skerry?

16. **Ru Clach Bane More (1).** NM 6222 7195. G. *Rubha na Cloiche Bàine Mòire, Rubha nan Clach Bàna Mòra* or *Rubha nan Clach Bàna Mòr* 'the greater *Rubha nan Clach Bàna*', or 'the headland of the big white rock'. Named *Rubha na Cailleach* 'the headland of the old woman' on OS maps.

17. **Ru Port an Da Salen (1).** G. *Rubha Port an Dà Shàilein* 'the headland of the port of two inlets'. Named 'Farquhar's Point' or *Rubha Mhic Artair* 'MacArthur's headland' on OS maps.

18. **Island Rhul (1).** NM 6285 7290. G. *Eilean Raghnaill* 'Ranald's island'. Named *Eilean Raonuill* on OS maps and described by Bald as belonging to Clanranald, so it probably formed part of the township of Bailetonach, Eilean Shona.

19. **Illan na Lusta (1).** NM 6425 6925. G. *Eilean ?* The OS has *Eileannan Loisgte* 'the parched islands' for the two small islands here, and the collected name may simply be a phonetic rendering of this.

20. **Ru Port a' Varra (1).** NM 6350 7254. G. *Rubha Port a' Bharra* 'the headland of *Port a' Bharra*'.

21. **Port Varra (1).** NM 6331 7246. G. *Port a' Bharra* 'the port of the hill'. The ground in the vicinity is unsuitable for cultivation, and the port is inconvenient for anyone living in any of the Ardtoe houses, so 'port of the crop' is unlikely.

22. **Core Island (1).** NM 6354 7270. G. *Còrr Eilean* 'the precipitous/uneven island'. Possibly derived from *Coire* 'cauldron' or 'whirlpool'?

23. **Glass Illan (1).** NM 6328 7279. G. *Glas Eilean* 'the blue/grey island'. Named *Sgeir an Dùrdain* 'the humming skerry' on OS maps.

24. **Skeir a Claiev (1). Sgeir na Claidheimh (2).** NM 6394 7235. G. *Sgeir a' Chlaidheimh* 'the skerry of the sword'.

25. **Islands Ruog (1).** NM 6410 7215. G. *Eileanan nan Ròg* 'the shag islands'. *Ròg* is a Gaelic loan-word from Old Norse *hrókr* 'shag', a seabird related to the cormorant. However, the documented form could simply be a part translation of G. *Na h-Eileanan Ruadh* 'the red islands'.

26. **Port Laach (1).** NM 6391 7210. G. *Port Làthaich* 'the muddy port', but Duncan Cameron said that locally it was 'the port of the quicksands'.

27. **Port Caulten (1).** NM 6360 7222. G. *Port C...?*

28. **Nuil Dou (1).** *NM 6450 7166.* Named *Faodhail Dhubh* 'the black sea ford', on OS maps, and the name recorded by William Bald in 1806/7 was simply a phonetic rendering of *An Fhadhail Dhubh*. The bay dries out at low water, leaving the remains of a tidal fish-trap exposed. G. *An Fhadhail Dhubh*, but it may have been pronounced *An Fhadhail Dubh*.

29. **Ru Vurr-n (1).** NM 6246 7100. G. ? *Rubha Bh...?*, 'the headland of the ...?'.

30. **Finlason's Bay (14).** NM 6445 6984.

31. **Waterfoot (14).** NM 6578 7111. G. *Bun na h-Abhann*. An alternative name for Shielfoot.

32. **Glac Ban Bheg (2).** NM 6250 7150. G. *A' Ghlac Bhàn Bheag* 'the small white valley'.

33. **Rudha Coire na Feorag (2).** NM 6220 7155. G. *Rubha Coire na Feòrag* 'the headland of the squirrel corrie', although it is an unlikely habitat for a Red Squirrel.

34. **Tom Buidhe (2).** NM 6285 7185. G. *An Tom Buidhe* 'the yellow hillock'. Named *Càrn Mòr* 'the big heap of stones' on OS maps.

35. **Tom a' Fimire (2).** NM 6291 7148. G. ?*Tom an Fhuamhair* 'the hill of the giant'.

36. **Ton Mhor (2). NM 6275 7120?** G. *An Tòn Mhòr* 'the big, blunt headland'….but not in this location. Possibly An *Tom Mhòr* 'the big hillock'. Named *Carn Mòr* 'the big cairn' on OS maps.

Ardtoe

37. **Ardtoe More (3). NM 6284 707072.** G. *Àird Tobha Mhòr* 'the greater township of Ardtoe'.

38. **Sgeir an Aodach (2). NM 6185 7124.** G. *Sgeir an Aodaich* 'the skerry of the sails' or 'of the clothes'. Named *Sgeir an Eididh* on OS maps. See No. 14, above.

39. **Rudhanen as a na Feorag (2). NM 6222 7154.** G. *Rubha Fàsadh nam Feòrag,* Named *Rubha Fasadh nam Feocullan* on OS maps. *Fasadh* 'dwelling', or *fàsadh* ' protuberance'. *Feòcallan* 'polecat' or *Feòrag* 'squirrel'. See No. 8, above.

40. **Carn Rock (2). NM 6311 7285.**

41. **Bogha Ruagh (2). NM 6349 7285.** G. *Am Bodha Ruadh* 'the red skerry' or 'the red sunk rock'.

42. **Eilean Carnagh (2). NM 6328 7278.** G. *An t-Eilean Càrnach* 'the island full of cairns'. Named *Sgeir an Dùrdain* 'the humming skerry' on OS maps

43. **Eilean Corra (2). NM 6352 7270.** G. ?*Eilean Còrr* 'the precipitous/uneven island'.

44. **Cnoc na Gardich (2). NM 6355 7168.** G.? *Cnoc na Gairdeach* 'the hill of the merrymaking'. Named *Cnoc nan Sguab* 'the hillock of the sheaf of corn ?' on OS maps. Was this the site of 'Harvest Home' celebrations at one time?

45. **Tigh Calum (28)**. G. *Taigh Chaluim* 'Calum's house'. Possibly named after Calum Domnhuil Cameron (?).

46. **Uamh an Fhuamhair (28). NM 6235 7197.** G. *Uamh an Fhuamhair* 'the giant's cave'.

47. **Loch Ceann Traigh (16). NM 6400 6900.** An alternative name for Kentra Bay.

48. **Ardtun Crofts, Kentra (40).** The Tobermory Court Records record the sudden death of Catherine MacPherson or McMillan, who died in childbirth in 1859, the delivery being supervised by James Howie, Surgeon to the Parochial Board since 1845 (40).

49. **Am Port Mòr (46). NM 6435 6996.** G. *Am Port Mòr* 'the big landing place'.

Names on Kintra Moss, held in common with the Shielfoot crofters.

50. **Port nam Baraillean (46). NM 6460 7086.** G. *Port nam Baraillean* 'the landing place of the barrels or casks'. On one of the Newton of Ardtoe crofts at the head of *An Fhadail Dubh,* and only operative when the tide is fully in.

51. **Cnoc nan Carna (46). NM 6490 6957.** G. *Cnoc nan Carna* 'the hillock of the cairns' or 'stones'. This hillock is on the route from Ardtoe to Eilean Fhianain and there are several coffin cairns here.

52. **Allt na Páirce Moire (46). NM 6505 7026.** G. *Allt na Páirce Moire* 'the burn of the big park (field or enclosure)'.

53. **Allt Uaimh Chaoruinn (46). NM 6495 6984.** G. *Allt Uaimh Chaoruinn* 'the burn of the cave (or hollow) of the rowan'.

54. **A' Pháirce Mhòr (46). NM 6510 6977.** G. *A' Pháirce Moire* 'the big park'.

55. **Dùn an Eirighe (46). NM 6468 6935.** G. *Dùn Eirighe* 'the fort of the chief', from *eirigh* 'viceroy, chief, governor'. Named *Dùn an Eididh* on OS maps, but this name appears to have been derived from the adjoining island, *Eilean an Eididh.* The fort is on a croft named 'Duneira', an anglicisation of the Gaelic name.

56. **Mointeach Mhòr Ceann Traigh (46). NM 6550 6950.** G. *Mointeach Mhòr Ceann Traigh* 'the great moss of Kentra'.

57. **Bagh Ceann Traigh (46). NM 6400 6900.** G. *Baigh Ceann Traigh* 'Kentra Bay'.

Vitrified wall of the fort *An Dùn,* Shielfoot

21

ARIVEGAIG : Àirigh Bheagaig : the small shieling ?

1807, Arrevegaig (1); 1828, Arevigag (24); 1847, Arivegaig (40); 1875, Arevegaig (8); 2002, Arivegaig (9).

Arivegaig, across the *Allt Beithe*

Valuation: Arivegaig does not appear in the early rent rolls or sasines, and it may have been a shieling of Acharacle until the Riddells bought the estate in 1767. In Murray of Stanhope's 'Anatomie of Ardnamoruchan and Swinard' of 1723 (7), he mentions 'shell sand' as one of the products of Acharacle, which suggests that it extended to the coast at that time, and the township boundaries as shown on Bald's Estate Plan of 1806-7 also suggest that this could have been the case.

The *Allt Beithe*

Sea-pool at the mouth of the *Allt Beithe*

Area: In 1807 the 807.07 acres contained 16.670 arable, 1.580 cultivated with the spade, 631.420 moor & pasture and 157.40 moss. Alex. Lowe wrote *'This is but a poor ill situated Possession, surrounded with moss'*.

Tenants: 1828, 7 tenants on a year to year basis (24); 1847, 10 crofters (40).

Settlements: Arivegaig, NM 6562 6816.
                   Aultbay,   NM 6538 6759.
                             NM 6645 6703

Shielings: Not recorded.

Placenames:
   1.   **Arivegaig (9). NM 6562 6816.** G. *Àirigh Bheagaig. Àirigh* 'shieling', but the specific *bheagaig* is unclear. Somerled MacMillan (38) translates *bheagaig* as 'of the little one', but what does this mean?

2.   Ary Bigg (14). NM 6648 6700. G. *Àirigh Beag* 'the small shieling'. Now referred to as **'Old Arivegaig'.**

Kentra Bay from 'Old Arivegaig'

The ford, along which the sea encroaches at very high tides, is shown running diagonally across the picture, and is bounded along the west side by the field dyke which encloses the lazy-beds

3. **Ru Illan Da Try (1). NM 6464 6824.** G. *Rubha Eilean Dà Thràigh* 'the headland of the island of two beaches'. The headland is converted into an island at spring tides, when the sea cuts across the neck of land which joins the two beaches. This route forms part of the *Faodhail a' Ghath,* and skirts the field shown on the Estate Plan of 1806, but not on the 1st Edition OS map surveyed in 1872.

4. **Faodhail a' Ghath (46). NM 6474 6824.** G. *Faodhail a' Ghath* 'the sea-ford of the javelin'. This is probably the ford where John Ogg, 9th (?) MacIain of Ardnamurchan was murdered by his uncle Angus Mòr (or Donald) MacIain about the year 1600. The ford runs between Kintra and Arivegaig.

5. **Rubha Faodhail a' Ghath (46). NM 6457 6832.** G. *Rubha Faodhail a' Ghath* 'the headland of the sea-ford of the javelin'.

6. **Illan na Shumraig Beg (1). NM 6427 6843.** G. *Eilean nan Seamrag Bheag* 'the little island of the shamrocks'; possibly Bird's-foot Trefoil.

7. **Aultbay (1). NM 6546 6760.** G. (*An t-*) *Allt Beithe* 'the birch-tree burn'.

8. **Ru Craig an Glass (1). NM 6485 6858.** G. *Rubha Chreagan Glasa* or *Rubha nan Creagan Glasa* 'the headland of the blue/grey rocks'.

9. **Skeir Dou (1). NM 6450 6828.** G. *An Sgeir Dhubh* 'the black skerry'.

10. **Ault Foilgach (14). NM 6524 6878.** G. *Allt Fadhail a' Ghatha* 'the burn of the sea ford of the arrow or javelin'. Named *Allt a' Ghatha* 'the burn of the javelin' on OS maps.

The house at 'Old Arivegaig'

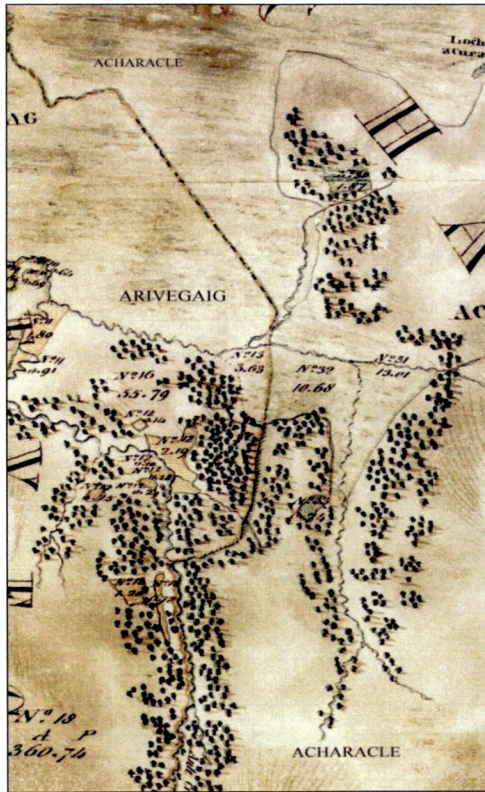
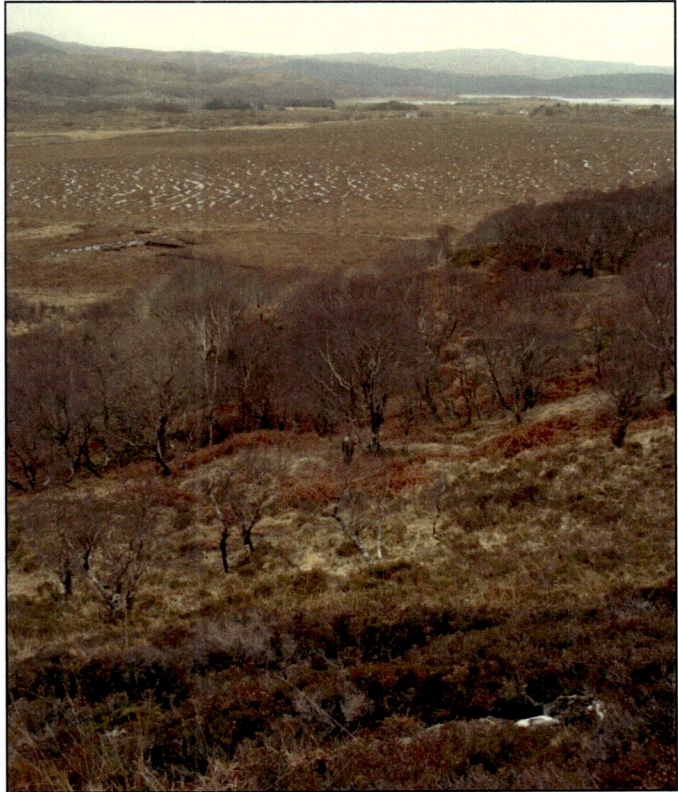

Old Arivegaig (detail)   National Records of Scotland,   Looking across the Dìg Bhàn to the Allt Beithe
RHP72/1-8, Plan of Ardnamurchan and Sunart, Argyll, 1806.

11. **Ault Foilbane (14); Faoil Vannan (1).** G. *Allt na Fadhail Bàine* 'the burn of the white sea ford'. For Faoil Vannan, cf. Ardtoe no. 2.

12. **An Fhaodhail (28). NM6562 6816.** G. *An Fhadhail* 'the sea ford'. The present site of Arivegaig was known by this name at one time.

13. **Pairce Mòr (28). NM 6504 6798.** G. *A' Phàirce Mhòr* 'the big field' or 'the great park'.

14. **Ath na feusgain (28), Àth na Fiasgan (46). NM 6445 6854.** G. *Àth nam Feusgan* 'the ford of the mussels'. The 1st Edition OS map shows the ford skirting an area labelled 'Mussel Scalps'. The ford runs between Arivegaig and Gorteneorn, joining the *Faoil na Fearna* immediately to the south of *Eilean nan Gad Beag*

15. **Creagan Glas (28). NM 6451 6832.** G. *Na Creagan Glasa* 'the grey rocks'.

16. **Faothal Dhonuill Chonullaich (36). NM 6524 6878** . G. *Fadhail Dhòmhnaill Chonallaich* 'Donald Connel's sea-ford'. Probably an alternative name for *Fadhail a' Ghatha.*

25

GORTENEORN : An Goirtean Eòrna: the little cornfield.

In 1541 and 1651, the 2½ merk land of Aldor, 6/8d land of Laik, 6/8d land of Longarie, and the 6/8d land of Craig were listed between Claish & Ardrimnish, and Ardtoe, suggesting that they should be included in Gorteneorn and Arivegaig (47, 11); 1667, Aldore, Lock, Langarie and Craigow (5); 1686, Gorton Zeorne (26); 1694, Gortanorm (6); 1722, Gorteneorn, Dale, Rineneach, Clachaig, Lungary and Craigdow (27); 1722, Leckick (27); 1723, Lecki(sh?) (10); 1723, Daal and Gortonoorn (7); 1737, Gorteneorn, Leakich (13); 1737, Dalle (26); 1784, Gortinforne (39); 1806, Arrein and Gortenean (19); 1824, Gorteneorn (24); 1875, Gorteneorn (8); 2002, Gorteneorn (9).

Gorteneorn from Ariean

Valuation: Laik, Longarie and Craig were each valued as 6/8d lands in 1541 and 1651 (47, 11) and as a 6 penny land in 1723, with Gorteneorn, Dale, Rineneach, Clachaig, Lungary and Craigdow each valued at 1d. Gortinforne and Craigdow formed a six penny land in 1784.

Area: Of the 3993.03 acres in 1806, 23.60 were arable, 21.18 cultivated with the spade, 3675.76 moor & pasture and 273.23 wood ground. It was let along with Glenborodale and Glenbeg to form a unit of 7297.39 acres.
This is a very extensive Farm. The greatest part except …high and rugged. Where there is a demand for Farms it may at first sight appear to some people to be desirable to cut it down into more possessions than one, but the Valuator has never advised cutting down or reducing the size of large stock farms if they be in the hands of a substantial Tenant. However, if the Managers of the Estate are of a different opinion it can be cut down with propriety only into two Farms. Glenboradale and Glenbegg making one, Gortenean and Arrein the other. To these last the access is extremely bad if cut off from that which passes over the Hill of Boradale to them. Alex. Lowe, 1807.

Tenants: 1686. Allan McEwin VcEan duy alias Cameron and his wife Mary; 1694, William Cameron with his wife Christian and children Donald, Christian, Moir, Mary and Elizabeth (6); 1716, Donald McAllan VcAlester, Allan McDhoil oig VcAlester, John McDhoil oig VcAlester, Allan McDhuil Soan alias Cameron, Alexander Cameron and his son Alexander, Donald McWilliam VcBean, Allan Cameron (workman), a vagabond (37); 1737, Alexander Dow Cameron (12); 1737, Ewn Cameron in Leakich (13); 1742, Alexander Dow mc Allester vic Dhouil alias Cameron in Dale; 1806, Capt. Alexander Cameron (19); 1828-1840, Alexander Cameron (24). 1864, 8 crofters and 1 householder were 'cleared' (43); 1870, 1 farmer and 3 crofters were 'cleared' (43).

Population: 1723, Daal and Gorteneorn, 8 families with 12 men, 12 women and 19 children, total 43.

26

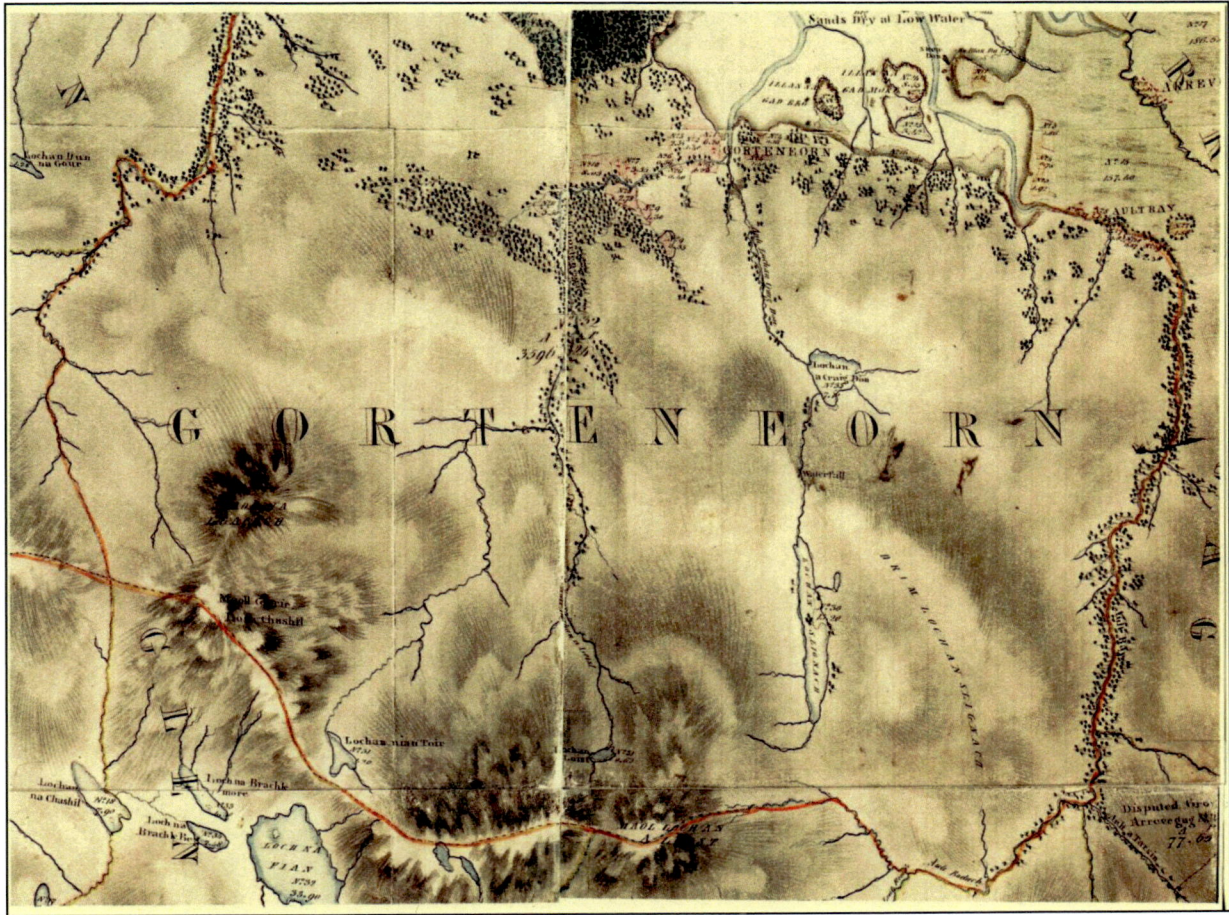

Gorteneorn (detail)   National Records of Scotland, RHP72/1-8, Plan of Ardnamurchan and Sunart, Argyll, 1806.

Settlements: NM 6330 6782. *Gorteneorn.*

NM 6550 6747. *'Torran Iamhair'.*

NM 6304 6980. *Camasdail.* House, barn and byre.

NM 6235 6874. *Daal of Kintra,* Possibly covered by windblown sand. Settled from the Mesolithic period onwards.

NM 6356 6773. *Ariean.* House, barn, byre and small animal pens.

NM 6100 6858. Leckish. House, barn, byre and kale patch, but much obscured by trees.

NM 6355 6685. *Creag Dhubh.* Field system and lazy-beds.

Shielings: NM 6266 6700. Ruidhe Breac. Developed shieling with a field system and lazy-beds.

NM 6553 6575.

NM 6465 6648. Àiridh Iain. Developed shieling with a field system and lazy-beds.

Placenames:

1.   **Gorteneorn (9). NM 6330 6793.** G. *An Goirtean Eòrna* 'the little cornfield'.

2.   **Ault Lochan a' Loist (1). NM 6269 6600.** G. *Allt an Lochain Loisgte* 'the burn of the burnt lochan' or more likely 'of the frog'. Named *Allt a' Ghoirtein –eòrna* 'the burn of the small barley field' on OS maps.

3.   **Maoll na Leacach (1). NM 6087 6592.** G. *Maol nan Leacach* 'the bare hill of the slabs', perhaps with gen. pl. of *leacach* 'place of slabs'. Named *Leac Shoilleir* 'the shining slabs' on OS maps.

4.   **Allt Lochan Creag Dou (1).** *NM 6365 6712.* G. *Allt Lochan Chreagan Dubha* 'the burn of the lochan of the black rocks'. Named *Allt Eas an Taileir* 'the burn of the tailor's waterfall' on OS maps.

27

5. **Drim Lochan Slignach (1)**. NM 6446 6540. G. *Druim Lochan Slignich* 'the ridge of the small shelly loch', but what kind of shells occur in a hill lochan? Probably with G. *Sligeanach* 'spotted, green, sky coloured'.

6. **Ault Bodach (1)**. NM 6460 6435. G. *Allt a' Bhodaich* 'the old man's burn'.

7. **Illan na Gad More (1)**. NM 6433 6815. G. *Eilean nan Gad Mòr* 'the big island of the withies'. Named *Eileanan nan Gad* 'the islands of the withies' on OS maps.

8. **Illan na Gad Beg (1)**. NM 6394 6802. G. *Eilean nan Gad Beag* 'the small island of the withies'.

9. **Eilean nan Gad Mollach (46)**. G. *Eilean nan Gad Mollach* 'the rough island of the withies'. An alternative name for *Eilean nan Gad Beag*, and included in *Eileanan nan Gad* 'the islands of the withies'.

10. **Skeir Torra Murchy (1)**. NM 6160 6834. G. *Sgeir Tòrra(idh) Mhurchaidh* 'Murdo's burial skerry'. Named *Sgeir an Rathaid* 'the skerry of the track' on OS maps.

11. **Port n'a Nian (1)**. NM 6180 6920. G. *Port nan Eun* 'the port of the birds'.

12. **Cual na Cross (1)**. NM 6195 6950. G. *Cùl na Croise* 'the back/nook of the cross'. The OS puts *Cùl na Croise* in the adjoining bay, ½ km to the NE.

13. **Ru Daal (1)**. NM 6259 7025. G. *Rubha nan Dail* 'the headland of the meadows'.

Camas Dail Bhuair from Ardtoe

14. **Camus Dall Vurr (1)**. NM 6304 6980. G.? *Camas Dail Bharra* 'the bay of the field of the hill'. The Camerons in the Dail of Kintra were famous rustlers and manuscript sources mention a good many stolen beasts being found there, so possibly with *bhuair* 'herd'.

15. **Ru Vurra (1)**. NM 6309 6982. G. *Rubha Bharra* 'the promontary of the hill' or possibly with *bhuair* 'herd'.

16. **Ru Taigh Store (1)**. NM 6359 6932. G?. *Rubha an Taigh Stòr* 'the promontory of the storehouse'.

17. **Illan na Shumraig (1)**. NM 6376 6896. G. *Eilean na Seamraig* 'the clover island'. Named *Sgeir na Seamraig Iosal* 'the lower skerry of the clover' on OS maps. *Seamraig* is a plant with trefoil leaves, like clover, shamrock, wood sorrel etc. In a situation like this, Bird's-foot Trefoil is the most likely.

18. **Ru McLeod (1)**. NM 6350 6890. G. *Rubha MhicLeòid* 'MacLeod's headland'.

19. **Kintra (1). NM 6300 6900.** G. *Ceann Tràigh* 'the head of the beach'. Bald named the peninsula to the west of Kintra Bay as 'Kintra'.

20. **Gorton nevin (10).** Was this an alternative name or confused spelling of Gorteneorn? The Rental in the reference names 'Leckish, Dale of Kintra, Gortennevin and Craiglow'.

21. **Kinira Bay (16). NM 6400 6900.** Should this have been Kintra Bay?

Drifting sand at Dail of Kintra

22. **Daal (1), Dale (27). NM 6235 6974.** G. *An Dail* 'the meadow' or 'field'. This settlement is named 'Dale of Kintra' in 18th and 19th century documents and the machar was occupied from Mesolithic times onward. Drifting sand has caused a lot of problems in the past. '*To the north of Gortenfern are some very large Hillocks of sand where they say some houses formerly stood but are now blown up' (11)...in 1767.* ie. Eroded or buried by shifting sand. This area was used as a firing range during WWII and unexploded ordnance still turns up in the dunes.

Ariean, a small settlement amongst coppiced oak.

23. **Ariean (17). NM 6356 6773.** G. *Àirigh Iain* 'Ian's shieling', or G. *Àirighean* 'the shielings'. See No. 34.

Creag Dhubh and a small enclosure across the lochan

24. **Craiglow (10), Craigdow (27). NM 6355 6685 (?)** G. ?*Creag Dhubh* 'the black crag'? A settlement valued as a 6/8d (= 1d land) in 1651. Named *Craigdow* in GD 241/64. The 6/9d land of **Crag** in 1610 (35). There is an extensive area of unenclosed and ill-defined lazybeds at this point, with *Creag Dhubh* rising precipitously to the south and east. **Crag** was lying waste in 1541.

25. **Rineneach (27).** G. *Ruighe nan Each* 'the shieling or grazing for horses'. A settlement probably valued as a 1d land in 1723. The location is uncertain. **Corrynenach,** a 1 merk land, was let to Rore McAlister in 1541 (47).

26. **Clachaig (27). NM 630 683** . G. *Clachaig* 'the stony place'. A settlement valued as a 1d land in 1723. Probably in the vicinity of *Camas Clachach*, but the precise location is uncertain.

27. **Droinen (29). NM 622 694.** G. *An Droighnean* 'the thicket of blackthorn'. When the houses at the Dale of Kintra were overwhelmed by the sand, the houses were said to have been built here, in the lee of a knoll. There is a burial ground nearby, but this has still to be located.

The jetty on Craig Earich

28. **Craig Earich (28). NM 631 683 ?** G. *A' Chreag Earraich* 'the spring(time) rock'. A prominent shoreline rock used for boarding boats, with a now ruinous jetty built alongside.

29. **Loch Ceann Traigh (16). NM 6400 6900.** Named 'Kintra Bay' on OS Maps.

30. **Caolas Ardtoe (16). NM 6270 7034.** 'Ardtoe narrows'.

31. **An traigh (16). NM 6350 6850.** G. *An Traigh* 'the beach'.

32. **Aldor (35).** Valued as a 2½ merk (= 5d) land in 1610, suggesting that it may have been an earlier name for Gorteneorn. It was lying waste in 1541 (47).

33. **Longary (34); Longarie, (35), Lungary (27, 46). NM 6291 6757.** G. *Longairidh* ON *Langærgi,* 'the long shieling', or 'the long vaccary'. An independent settlement valued at 6/8d, a 1d. land in 1541 and 1651, but later included in Gorteneorn. The 1st Edition OS map shows a fank here, but this has since been removed. Longary was lying waste in 1541.

Lazybeds on Àiridh Iain and the view across Lochan na Creag Duibhe to the crags of Creag Dhubh

34. **Àiridh Iain. NM 6465 6648.** G. *Àiridh Iain* 'Iain's Shieling'. John Cameron 'Raeland' suggested that this was the correct name for Àird Iain at NM 6433 6604 (46), but there is no shieling here, and besides, the *Allt Àiridh Iain* is at NM 6460 6638, with a developed shieling nearby, consisting of a field system, buildings and a sheepfank at NM 6490 6657. At a later date, the name appears to have been transferred to NM 6356 6773, 'Ariean'. Cf. no. 21.

35. **Ruidh Breac (46). NM 6262 6701.** G. *Ruidhe Breac* 'the brindled grazing ground or shieling'. An exensive area of superior pasture with a sheep fank at the quoted grid reference. Was this *Coire/Ruighe nan Each*, the 6/8d land of 1541-1723?

36. **Torr a' Chuillinn (46). NM 6495 6660.** G. *Tòrr a' Chuilinn* 'the hill of the holly'.

37. **Torr Garbh (46). NM 6556 6729).** G. *Tòrr Garbh* 'the rough hill'.

38. **A' Mhaorach (46). NM 6450 6790.** G. *A' Mhaorach* ' the place where shell-fish is found', from *maorach* 'shell-fish in general, mussel, limpet, fishing bait or a place where shell-fish are found'.

39. **Leckish (10). NM 6100 6858.** ?G. *Leacaich* 'paved with slabs', but if so, then the place is not well named. **Laich** was lying waste in 1541. The 6/8d land of **Laith** in 1610 (35).

40. **Faoil na Fearna NM 6400 6790.** G. Faoil na Fearna 'the sea-ford of the alder tree'.

The barn at Leckish

Eilean nan Gad Beag (left) and the Faoil na Fearna at high tide

Eilean nan Gad Mòr from Rubh Eilean Dà Thràigh on Arivegaig

Inlet, lazybeds and a field dyke on *Eilean nan Gad Mòr*

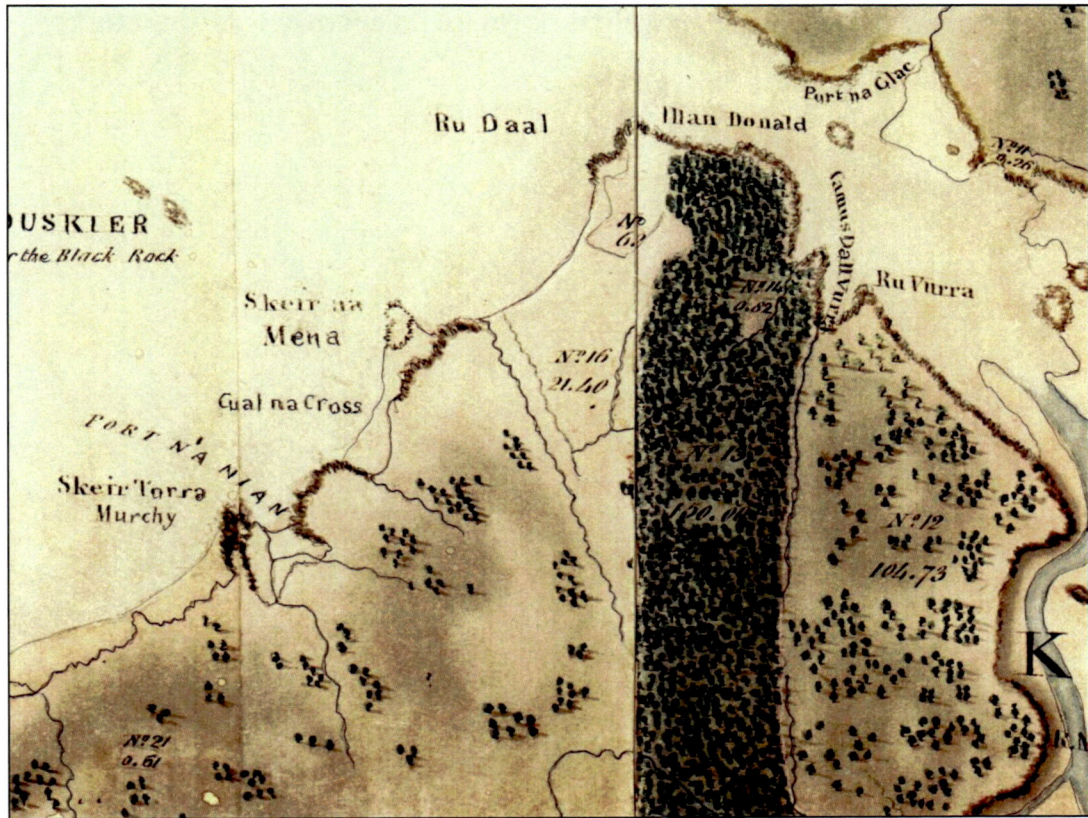

Gorteneorn (detail). Dail and the 'Singing Sands'.
National Records of Scotland, RHP72/1-8, Plan of Ardnamurchan and Sunart, Argyll, 1806.

A recessed platform below *Airidh Iain* and overlooking *Lochan na Creige Dhuibhe,*
possibly a shieling hut from the mediaeval period.

GORTENFERN : An Goirtean Feàrna : the little cornfield of the alder.

1541, Clas; Ardrenynes (47); 1610, Clas and Ardrenynes (35); 1651, Class & Ardrumneis (11); 1667, Classe & Ardyna (5); 1723, Clash and Ardrimonish (10); 1737, Gortenfearn and Ardriminish (22); 1784, the four penny land commonly called Peidow or Blackpenny (39); 1828, Gortnfern (24); 1841, Gortinfern and Craigdow (23); 1875, Gortenfern (8); 2002, Gortenfern (9).

Lochan na Clais   Photo: Owain J. Kirby

Valuation: In 1541 and 1651, Claish and Ardrimnish were each valued as 6/8d lands, and in 1723 jointly as a 2 penny land, and Gortonfern and Lehick were also valued as 1 penny lands each. In 1841, Gortinfern and Craigdow formed a six penny land (23), but this sounds like a misprint for Gorteneorn. The four black pennies were Claish, Ardriminish, Gortenfern and Liath Doire (or Garbhallt).

Area; In 1807 the 1542.439 acres consisted of 14.760 arable, 24.120 cultivated with the spade and 1506.559 of moor & pasture, and Alex. Lowe wrote *'This is neither a good or a pleasant farm, and it is almost inaccessible on all sides, except by the sea'.*

Tenants: 1541, Johne McErenoch in Clas; Ardrenynes was lying waste (47); , Alexander Cameron, Donald McEun ruay, John roy McInnish VcDhoil and John roy McKenrig (37); 1737, Alexander Cameron (22); 1738, Evan Cameron, Tacksman (12); 1741, Alexr Breachk Cameron, broyr to Evan Cameron (12); 1742, John mc Alister vic Dhouil vic Dhouil vic mc olony, Eun mc Dhouil alias Cameron, Alester mc Allen vic Allester vic Dhuil (23); 1744, Archd. McDiarmid (32); 1806, Alex. Cameron (19); 1828, Ritchie (24).

Population: Gortonfern and Lehick,1723, 3 families, with 5 men, 5 women and 4 children, total 14. Clash and Ardrimonish, 1723, 2 families with 4 men, 7 women and 2 children, total 13. 1851, nine cottar families with a total cottar population of 46 (44).

Settlements: NM 6030 6885. *Clash,*
                  NM 6074 6918. *Camas an Lighe,*
                  NM 5947 7098. *Liath Doire,*
                  NM 5839 7255. *Acarsaid Bheag*
                  NM 5870 7211 *Acarsaid Mor*
                  NM 5991 7043. *Garvalt,*
                  NM 6074 6887. *Gortenfern*

Shielings: Not known.

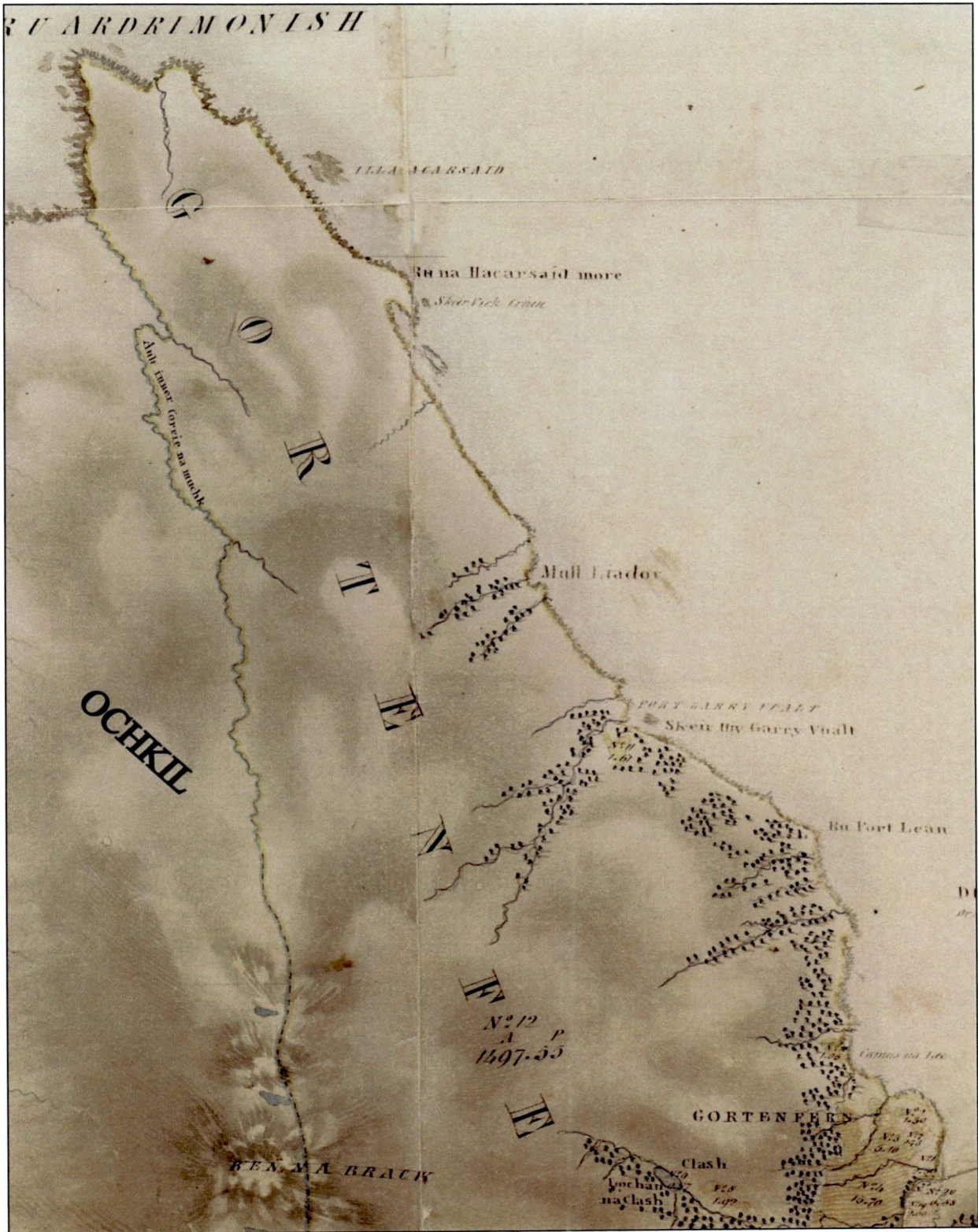

Gortenfern (Detail)   National Records of Scotland, RHP72/1-8, Plan of Ardnamurchan and Sunart, Argyll, 1806.

Placenames:

1. **Gortenfern (9). NM 6074 6887.** G. *An Goirtean Feàrna* 'the little cornfield of the alder'. This was part of the wadset lands held by Alexander Campbell of Ardslignish and described as *'the pennyland of Gortenfern in Kientrae, commonly call'd one of the four black pennies'*.

D-shaped lime kiln with a stepped-down lintel at Gortenfearn

2. **Ault inner Corrie na muchk (1). NM 5827 7132.** ? G. *Allt Inbhir Coire nam Muc* 'the burn at the mouth of the corrie of the pigs'. Named *Allt an Rudha Ruaidh* 'the burn of the red promontory' on OS maps. This was the march burn between Gortenfern and Ochkil.

3. **Skeir Vick Crian (1). NM 5912 7188.** G, *Sgeir Mhic C...?* A skerry followed by a personal name.

4. **Mull Liador (1). NM 5952 7116.** G. *Maol an Liath Doire* 'the hill of the grey grove', or possibly G. *Maol an Leathaid Òir* 'the hill of the golden slope'.

5. **Camus na Lee (1). NM 6080 6924.** G. *Camas na Lighe* 'the bay of the flood'. Named *Camas Liath* 'the grey bay' or *Camas an Lighe* on OS maps, and the name has moved slightly to the east at NM 6120 6920.

6. **Lochan na Clash (1). NM 6020 6882.** G. *Lochan na Clais* 'the lochan of the ditch'. Named *Lochan na Glaice* on the 1st Edition OS map.

7. **Camas ne Leigh (14). NM 6074 6918.** G. *Camas na Lighe* 'the bay of the flood'. A settlement that takes its name from the bay.

8. **An Innis Dubh (21).** G. *An Innis Dubh* 'the black (or second rate?) water-meadow'.

9. **Peendow or Blackpenny (3).** G. A' *Pheighinn Dubh/Dhubh* 'the black (or second rate ?) pennyland'. This is the four penny lands of Clash and Ardrimonish, commonly called 'Peindow or Blackpenny' (39).

10. **Liathadair (20). NM 5947 7098.** G. *Liath Dhoire* 'the grey grove'. A small settlement with several buildings scattered amongst the rocks.

Quay and cobbled working platform from the boatshed in Camas na Lighe

11. **Clash (1, 27). NM 6030 6885.** G. *A' Clais* 'the ditch' or 'the furrow'. There was probably a settlement here at one time but the remains have still to be located. In 1541 the assedatur was Johne McErenoch at a rent of 5 stones of meal. 5 stones of cheese and 2/6d in silver (47).

12. **Lochan Dun na Gour (1). NM 6000 6760.** G. *Lochan Dubh nan Gobhar* 'the black lochan of the goats'. Named *Lochan Dubh* 'the black lochan' on OS maps

13. **Ru na Hacarsaid More (1). NM 5905 7208.** G. *Rubha na h-Acarsaid Mòire* 'the headland of the big harbour' (or anchorage).

14. **Thun na Ranna (20).** The Gaelic is obscure and the location is not known.

15. **Rathad nam Mealla Ruadh (20).** G. *Rathad nam Meall Ruadh* 'the road of the bare red hill'. Described as 'Steep rocks on the south side *thun na Ranna* (North side of Ardnamurchan)'.

16. **Ardrimonish (1, 27), Ardrumneis (11). NM 5839 7255; NM 5870 7211.** Perhaps G. *Àird Riminish,* but it is uncertain if the specific is ''Driminish' or 'Riminish'. 18th and 19th Century documents refer to Gortenfern under several different names, including 'Clash and Ard Drimnish'. Ard Riminish is probably an earlier name for Acarsaid Bheag/Mór, a safe anchorage for vessels waiting for the weather to improve before rounding Ardnamurchan Point….and possibly a good location for a Viking beach market ? It was lying waste in 1541.

17. **Garvalt (8). NM 5991 7043.** G. An *Garbh Allt* 'the rough burn'. A small settlement that took its name from the adjoining burn. The port here is '**Port Garry Vualt**'

18. **Skeir Uiy Garry Vualt (1). NM 60 70.** G. *Sgeir Bhuidhe a' Gharbh Uillt* 'the yellow skerry of the rough burn'.

19. **Bathaich na Caillich (28). NM 5916 6979 ?** G. *Bàthaich na Caillich* 'the byre of the old woman'.

20. **Tobhar Lochan na Glaise (28). NM 602 688 ?** G. *Tobar Lochan na Clais* 'the well of the lochan of the furrow'.

21. **Lubailean (29). NM 5916 7171.** G. *Lùb Ailein* 'Allan's pool' or 'Allan's creek'.

22. **Meall na Leachach (16). NM 5880 6915.** G. *Meall na Leacaich* 'the hill of the slabs'. Named *Beinn Bhreac* 'the speckled mountain' on OS maps.

Lochan na Clais from the settlement

SALEN : Sàilean Loch Shuaineirt : the shallow inlet of Loch Sunart
1807, Salen (1); 1824, Salen Tarbert (41); 1828, Salen, (24); 1875, Salen (8); 2002, Salen (9).

Salen Jetty, built c. 1830 by the British Herring Fisheries Society, although the herring left Loch Sunart in 1818.

Tenants: 1816, Duncan Livingston, tenant; 1817, John MacNaughton Jnr; Donald MacNaughton, weaver; 1822, James MacNaughton, tenant; Donald Cameron, tenant; Duncan MacMillan, crofter (41); 1828-1842, Dugald McNaughton; 1864; the widow of the tenant was 'cleared' from the farm and Inn (42).

Settlements: NM 687 651. House and barn, with lynchets and rig.
NM 6887 6471. Salen. Salen does not appear in the early records and was probably a pendicle of Tarbert until the late 18th century.

Shielings: The township was too small to require shielings.

Placenames:

1. **Salen (9). Salen Loch Sunart.** G. (1), *An Sàilean; (2) Sàilean Loch Shuaineirt* 'the shallow inlet of Loch Sunart'. *Sàilean* ' little inlet or arm of the sea; a deep bay.

2. **Mill Stream (24). NM 6881 6460.** G. *Allt a' Mhuilinn.* This was let to John Clark along with the bobbin mill, mill dam and pond.

3. **Lochan a' Mhuilinn (46),** G. *Lochan a' Mhuilinn* 'the lochan of the mill'. **Salen Mill Dam (8). NM 6853 6477.** A large shallow lochan built to supply a head of water for the water-powered Salen Bobbin Mill. Water entering the mill pool came from *Lochan na Dubh-Leitir*, and was regulated by a dam and sluice at NM 6765 6526, and from *Loch a' Chaisil*, with the dam and sluice at NM 6700 6492.

4. **Sg. Ruadh (25). NM 6890 6436.** G. *An Sgeir Ruadh* 'the red skerry'.

5. **am Bréum Torr (20).** Described as being 'west of Salen'.

6. **Salen of Sunart (20) NM 6900 6400.** G. S*àilean Loch Shuaineirt.* Named **Tarbat Bay** on General Roy's Map of c. 1750, Salen Bay on the 1st Edition OS map and *Bàgh an t-Sallein* on current maps.

7. **Pirn Mill (5). NM 6884 6460.** Salen Bobbin Mill. This was built in 1840 and rented out to John Clark Junior of Mile-end, Glasgow, producing thousands of bobbins for use in Clark's cotton mills in and about Glasgow. Cutting and transporting locally grown birch and larch poles provided a lot of employment locally, but unfortunately it burned down in July, 1854, and was never rebuilt.

8. **Salen Auction Mart. NM 6883 6491.**

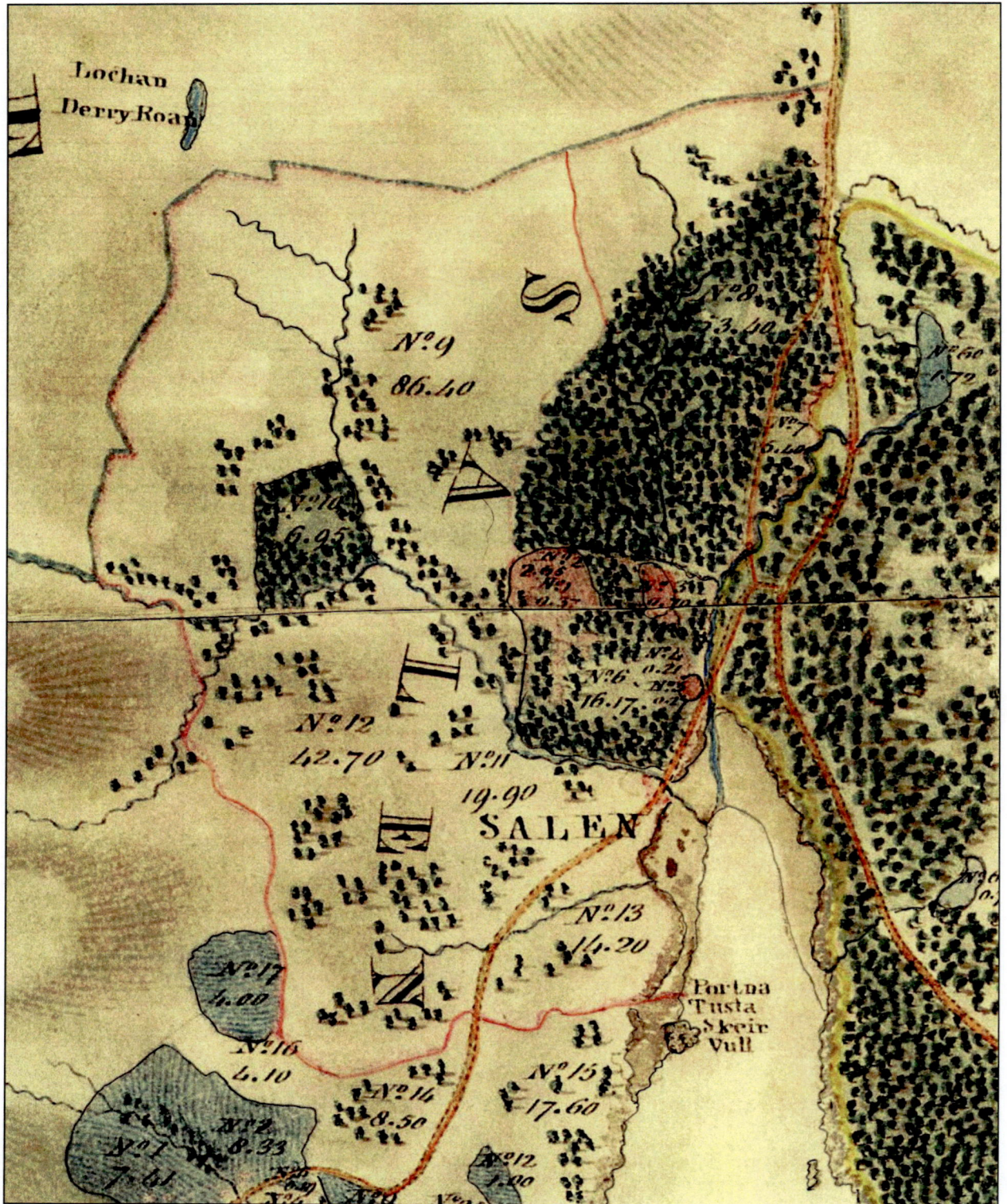

Salen

National Records of Scotland, RHP72/1-8, Plan of Ardnamurchan and Sunart, Argyll, 1806.

SHIELFOOT : Bun na h-Aibhne Seilich : the foot of the River Shiel
1723, Waterfoot (7); 1807, Sheil Foot (1); 1828, Sheilfoot (24); 1875, Shielfoot (8); 2002, Shielfoot (9).

Houses on Shielfoot

Valuation: Valued along with Ardtoe as a 2 merk land in 1667 and as a 4 penny land in 1723.

Area: The 314.570 acres consisted of 6.245 arable, 27.660 cultivated with the spade, 245.335 moor & pasture and 35.330 wood ground, plus access to the 533.07 acres of moss, held in common with Ardtoe.

On the North side of the Bay is a large Moss which they have been endiavouring to drain, but to no purpose as the Drains are all fill'd up & the Moss in many places impossible even for people on foot – betwixt which moss and the water of Shiel is another large cluster of Rocks covered with wood as above in the midst of it is a small improvement of about 2 acres belonging to the former Town all the Houses of the three last mentioned Towns good for nothing (18).

Tenants: 1807, Alex. Cameron and six others; 1828, 6 tenants on a year to year basis; 1847, 9 tenants (40).

Population: 1723, 3 families, with 3 men, 3 women, 5 children, total 11.

Settlements: *Shielfoot,* NM 6577 7111. 'Lower Shiel-foot'. At least seven houses with barns, byres, kale patches, animal pens etc. Corn kiln and the site of a church.
Shielfoot, NM 6592 7030 to NM 6635 6967. Modern crofting community.
Toragaltromen, NM 6630 7001. House, barn and byre c.1825.

Shielings: The township was too small to require shielings.

41

Shiel Foot. National Records of Scotland, RHP72/1-8, Plan of Ardnamurchan and Sunart, Argyll, 1806.

Shielfoot and the moss with Newton of Ardtoe in the distance

Placenames:

1. **Shielfoot (9). NM 6577 7111.** G. *Bun na h-Aibhne Seilich* 'the foot of the River Shiel'. The OS maps have for G. *Bun Abhainn Sheileach.*

2. **Port Ru na Nuan (1). NM 6488 7184.** G. *Port Rubha nan Uan* 'the port of the headland of the birds'.

3. **Skeir na Druiden (1). NM 6502 7222.** G. *Sgeir nan Druidean* 'the skerry of the starlings'…but it is an odd place for starlings.

4. **Ru P. Min (?) (1). NM 6474 7199.** G.? *Rubha a' Phuirt Mhin* 'the headland of the smooth port'.

5. **Port Bain (1). NM 6491 7190.** G. *Am Port Bàn* 'the white port'.

6. **Ru Dou Clan Nuil (1). NM 64547131.** G. ? *Rubha Dubh Chlann Nèill* 'the black headland of Clan Neil'.

7. **Cnoc na Glaic Salleich (2). NM 6483 7142.** G. *Cnoc na Glaic Seilich* 'the hill of the willow hollow'. Named *Meall Bun na h-Aibhne* 'the lumpy hill at the foot of the river' on OS maps.

8. **Ardtoebeg (?) NM 6578 7113.** G. *Àird Tobha Bheag* 'the lesser township of Ardtoe'. Known locally as 'Old Shielfoot'.

9. **Penbunochan (16). NM 6575 7137.** G. *Peighinn a' Bhunachain* 'the pennyland at the foot of the river'. Probably an earlier name for 'Old Shielfoot', and a one pennyland of Ardtoe.

10. **Waterfoot (7, 14). NM 6578 7113.** G. *Bun na h-Aibhne* 'the foot of the river'. An alternative name for 'Old Shielfoot'.

11. **Sgeir na Smeorach (2). NM 6500 7221.** G. *Sgeir nan Smeòrach* 'the skerry of the thrushes'. Possibly G. *Smeurach* 'full of bramble-berries', but brambles are as unlikely as thrushes here. Named *Sgeir nan Ròn* 'the skerry of the seals' on OS maps.

12. **Cnoc na h-Eaglish (2). NM 6553 7143.** G. *Cnoc na h-Eaglaise* 'the hillock of the church', from Latin: *Ecclesia.* In 1728, the Rev. James Stevinson complained about his work-load, having 3 places of worship in Sunart and 6 in Ardnamurchan, including Kintra. Services continued to be held here until about 1915.

13. **Port Eabar, (2). NM 6490 7182.** G. *Am Port Eabair* 'the slimy or muddy port'. Named *Port Bàn,* 'the white or fair port' on OS maps.

14. **Rudha Porst an Eabar (2). NM 6475 7194.** G. *Rubha Port an Eabair* 'the headland of the muddy port'.

15. **Durre na Achlaish (2). NM 6500 7186.** G. *Doire na h-Achlaise* 'the grove of the armpit'.

Eileann Uaine ; probably Inchbord, at the mouth of the river.

16. **Inchbord (20). NM 6595 7160.** G. *Iubhard.* 'The island at the <u>mouth</u> of Loch Shiel should be *Iubhard'.* Where is the island at the foot of Loch Shiel? Yet Cowley's Plan of 1734 shows 'The Salmon Leap' at the '<u>Mouth</u> of Loch Sheil'. Possibly with G. *Innse* 'island' and an Old Norse loan-name *–fjörðr,* 'fiord'.

17. **The Salmon Leap (14). NM 6598 7090.** G. *Eas Bun na h-Aibhne* 'the waterfall at the foot of the river'. A series of rapids and waterfalls at the foot of the River Shiel, navigable by small vessels at high tide.

18. **Cruach (2). NM 6550 7124.** G. A' *Chruach* 'the bold hill'.

19. **Sgiath Bheag (28). NM 6633 6994.** G. *An Sgiath Bheag* 'the small wing'.

20. **Tòrr famhair (41). NM 6630 7001.** G. *Tòrr an Fhuamhair* 'the giant's mound' or 'hill'. An alternative name for The Tòrr, Shielfoot. C. 1910, but not recorded elsewhere.

21. **Lùib (31). NM 6650 6920.** G. *Lùib* 'the bend'. The name of a house built near a bend of the river. Originally there were four turf houses here, built by people who moved from **Lower Shielfoot Strath** after many of the residents there died of the 'black fever'.

22. **Na Carnaichean (46). NM 6595 7116.** G. *Na Carnaichean* 'the place of cairns'. Named 'Cairn Pool' on OS maps. A tidal fishing pool at the foot of the River Shiel.

44

Shiel-foot and the South Channel from
Bealach Sgairt Dea-uisge in Moidart

The moss in many places impossible
even for people on foot

23. **Toragaltromen (14). Torra alt ra nian (1). NM 6630 7001.** There is no *allt*, 'burn' here apart from a trickle of water coming from springs/wells on either side of the field. G. Possibly *Tòrr an Altramain* 'the hill of the nursling'. A small settlement between the two rocky, tree-clad summits of The Torr, Shielfoot.

Am Bioda Dubh and The Torr, Shiel-foot

The house on Toragaltromen

24. **Am Bida (31, 46); Am Bioda Dubh (46). NM 6624 7023.** G. *Am Bida* 'the pinnacle'; *Am Bioda Dubh* 'the black pinnacle', 'the black pointed hill top'. The rocky summit of 'The Torr', Shielfoot, especially when seen from the NW. The summit is crowned by a vitrified roundhouse or *dùn* on the tip of a spectacular elongated vitrified fort which occupies the summit of the ridge.

25. **Coille an Torr (46). NM 6645 7013.** G. *Coille an Torr* 'the wood of the hill', 'Torr Wood'. A remarkable oakwood with dwarf wind-sculptured trees on the western side and larger, coppiced and pollarded trees on the more sheltered eastern side. '

TARBERT : An Tairbeart : the crossing.

1541, Tarbart (47); 1610, Tarbat, Tarbert (35); 1651, Camustorsay, Tarbert (4); 1667, Camstorsey, Tarbet (5); 1723, Tarbert, Camistorsay (6); 1723, Terbart (10); 1723, Camistorsa (7); 1740, Tarbett (26); 1740, Tarbert (12); 1742, Tarbet (12); 1750, Camistorse; Tarbat (8); 1784, Tarbert (23); 1828, Tarbert (24); 2002, Tarbert; Camastorsa (9).

Houses and 'offices' on Tarbert

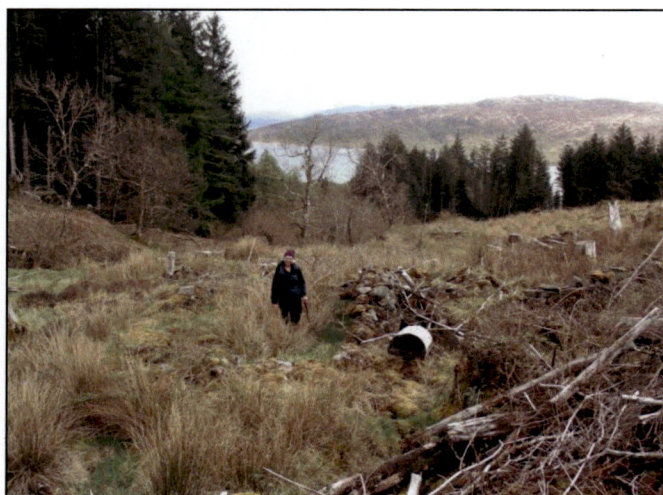

Valuation: Valued as a 1 merk land in 1541 and in 1610, and a 3 penny land along with Camastorsa in 1723. A two penny land in 1784 when Camastorsa was joined to Camustynnes to form a 1½ penny land (23).

Tenants: 1716, John Cameron, Sorle McColl, Donald Mcpherson, Alexander Mcphaill, Lachlan McOlony and his son John (37); 1740, Alexander Cameron, tacksman (12); 1742, Eun mc Pherson (12); 1828, 2 tenants, with D McKinnon in the croft (24); 1851, Three brothers, Duncan, Donald and John Cameron; 1847, 1 tenant and 2 crofters (40); 1856, Hugh Cameron, commonly called 'The Cannon' in Camustorsay; James Cameron, son of John Cameron, farmer, Tarbet (40); 1864, the 3 farms were 'cleared, as was the Camastorsa croft (43).

Population: 1723, 4 families, with 6 men, 6 women, 10 children, total 22.
            1851, 3 families, total 24.
Settlements: NM 6824 6378. *Tarbert,*
            NM 6710 6216. *Camas Torsa,*

Shielings: Not known.

Between
rbet

Lochan cashi
Nº 21
7.80

T
A
R
B
E
R
T

Nº 26
A P
793.14

Lochan
Timaemind

TARBERT

Nº 18
5.75

Nº 10
73.80

Nº 20
43.70

RU AN TORMOLLOCH

Illan Irach

CAMUS
Nº 10
A R
60.00

Ru Suna Port

SALEN

Nº 12
42.70

Nº 11
19.90

Nº 13
4.20

Nº 17
4.00

Nº 16
4.10

Nº 16
8.50

Nº 14
17.60

Nº 15
17.60

Nº 12
1.00

Portna
Tusta
skeir
Vult

Ru na nish
Ruey

Ru Vult

Port na Batachin

Ru Vichk Mc Lean

H

Ru na Coille Darrach

Tarbert (detail).   National Records of Scotland, RHP72/1-8, Plan of Ardnamurchan and Sunart, Argyll, 1806.

Placenames:

1. **Tarbert (9). NM 6824 6378.** G. *An Tairbeart* 'the crossing' or 'the portage'. The township took its name from the portage between Loch Sunart and Loch Shiel, but by about 1800, the NW portion was detached to form Salen. In 1541 the rent was vj bollis j ferlot j pec malt the thrid of ane mart and thrid part of the half quarter of ane mart x stane cheis x stane mele (47).

2. **Camastorsa (9). Camus Torsay (1). NM 6710 6216.** G. *Camas Tursa* 'the bay of the giant' or 'monolith'; or *Camas Tùrsa/Camas Tùirse* 'the bay of mourning'. It is unlikely to incorporate the Old Norse word *á* ' a river', as there isn't even a burn worthy of the name in the vicinity.

3. **Ault Tarsin (1). NM 6585 6446.** G. *An t-Allt Tarsainn* 'the transverse burn'. Named *Allt Beithe* 'the birch burn' on OS maps.

4. **Skeir Vull (1). NM 6891 5413.** G. *? An Sgeir Mhaol* 'the bare skerry'.

5. **Ru na nish Ruay (1). NM 6880 6380.** G. *? Rubha na h-Innse Ruaidhe* ' the headland of the red meadow', with *Ruaidhe,* a compound of *ruadh* ' red'.

Port Rubha a' Mhuilt

6. **Port Ru Vuilt (1). NM 6866 6354.** G. *? Port Rubha a' Mhuilt* 'the boat landing of the headland of the wether' (young male sheep).

7. **Ru na Trivan (1). NM 6857 6344.** G. *Rubha na Tràigh Bàine* 'the point of the white beach'.

8. **Port na Batachin (1). NM 6845 6349.** G. *Port nam Batachan* or *Port nam Bàtaichean* 'the landing place of the small boats'.

9. **Port Vichk McLean (1). NM 6821 6324.** G. *Port MhicGill-Eain* 'MacLean's landing place'.

10. **Ru an Tormolloch (1). NM 6780 6246.** G. *? Rubha an Torra Mholaich* 'the headland of the rough hill'.

11. **Illan Irach (1). NM 6746 6217.** G. *An t-Eilean Ìorach* an island name, possibly with *iorrach* gen. *iorraich* 'quiet, undisturbed'.

12. Ru Suna Port (1). NM 6660 5128. G. *Rubha* ...? A headland name with a boat-landing adjacent.

Rubha na Tràigh Bàine and Port nam Batachan

13. Tom na Buachailleart (2). NM 6720 5342. G.? *Tom na Buachailleachd* 'the hill of the hearding'? Probably a derivative of *buachaill* 'cow-herd, shepherd or one that watches over cattle'.

14. Goirtean Gun fhios (3). G. *An Goirtean Gun Fhios* 'the unknown little cornfield', or possibly more plausibly *Gun Fhiùbhas* 'without worth'. Described as being '1½ miles W of Salen'. There are several small enclosures to the east of the Camas Torsa crofts and it is uncertain which one this name refers to.

15. Point of Tarbert (11). NM 6880 6353. G. *Rubha an Tairbeirt* 'the headland of the crossing' or 'Tarbert Point'. This is, presumably, *Rudha Bhuailte* on OS maps.

16. Lochan Timaguind (1). NM 6632 6319. G. *Lochan* ...?

17, Port na Tusta (1). NM 6886 6422. G. *Port na T...?* The name of a boat landing.

17. Allt an Tairbeirt (46). NM 6831 6379. G. *Allt an Tairbeirt* 'the Tarbert burn'.

18. Illan Irach (1). Eilean Aighearach. NM 6744 6220. G. *Eilean Aighearach* 'the festive island' from *aighearach* 'glad, mirthful, joyful, gay, festive'.

REFERENCES

1. NRS. RHP 72, 1-8. William Bald's Estate Plan of the Barony of Ardnamurchan and Sunart, (1806).
2. Admiralty Chart No. 531. *Loch Moidart*, 1860. Surveyed by James Jeffrey Master R.N. with additions by Capt. E.J.Bedford, R.N. in 1865.
3. NRS, GD 241 Box 164, 2,3. Alexander Bruce's *Plan of Loch Sunart (1733)*.
4. NRS. AF 49/4. Riddell Mss. Ardnamurchan and Sunart Rentals, Charge and Discharge, copy leases etc
5. Registrum Magni Sigilli Regnum Scotorum: Register of the Great Seal of Scotland (RMS) Vol. XI. No.1105.
6. Grant, Francis J. Editor (1909). *The Commissariot of Argyll, Register of Inventories, 1693-1709.* The Scottish Record Society, Part XLIV. Skinner and Co, Edinburgh.
7. Murray, Sir Alexander, of Stanhope (1740). *The True Interests of Great Britain, Ireland and Our Plantations.* Appendix entitled 'Anatomie of the Parish and Barony of Ardnamoruchan and Swinard 1723'.
8. Ordnance Survey, 1st Edition Maps
   Argyllshire, Sheet XV, Surveyed 1872, published 1875;
   Argyllshire, Sheet XVI, Surveyed in 1872 by Capt. Bolland, RE, published 1875.
   Inverness-shire, Sheet CXLVII, Surveyed by Capt. Coddington RE in 1873, published 1876.
9. Ordnance Survey, 1:25,000 Scale, Explorer Map 390, *Ardnamurchan, Moidart, Sunart and Loch Shiel,* 2002.
10. NLS, Adv. Ms 29/1/1/ Vol.VII f.139-43. Murray of Stanhope Papers.
11. Campbell, Herbert (1933). Argyll Transcripts, Abstracts of the Argyll Sasines, Vol. II (1st Series) No.344, 1651. Edinburgh.
12. NRS. SC54/2/17/65/6. Summons of MacOlony Camerons and others for theft of lead from Mingary Castle in 1740 by Campbell of Lochnell. Inveraray Court Records.
13. NRS. C2/2/30/2. Summons by the Duke of Argyll to Debtors in Ardnamurchan, 1737.
14. I. Cowley (1734). *A Map of the Improveable Bay of Kintra drawn from a Plan of the Survey of Ardnamorchan.* London, 1734.
15. NRS. CC2/2/35/5. Summons by John Richardson, Factor, to sundry tenants in Ardnamurchan, 1739.
16. John Thomson's Atlas of Scotland No. B 10, 1832, North Argyll. Edinburgh.
17. CL/A9/3. George Langland's Map of North Argyll, 1801.
18. Unsourced transcript.
19. NLS. AF 49/2/ Rental of the Lands of Ardnamurchan and Sunart, 1806.
20. NLS Ms 369. Place-names of Ardnamurchan, Sunart etc. Probably Charles M. Robertson c.1900.
21. Henderson, Angus (1914-16), 'Ardnamurchan Place-names' in *The Celtic Review* **10,** 1914-16 pp. 149-168.
22. NRS. CC2/13/2.
23. NRS. GD 241/165. Riddell Papers, vouchers for expenditure.
24. NLS. AF 49/3. Thomas Anderson, 1 January, 1829. Report on the State of Farms on the Barony of Ardnamurchan and Sunart.
25. Admiralty Chart No.3185. *Loch Sunart.* 1864. Surveyed by Capt. E J Bedford RN.
26. Grant, Francis J. Editor (1902). *The Commissariot Record of Argyle; Register of Testaments, 1674-1800.* Scottish Record Society, Edinburgh.
27. NLS. Adv. Ms 29/1/1 Vol.VII, 133-136. Rental of Ardnamurchan and Sunart payable to Sir Duncan Campbell of Lochnell, 1722.
28. Mary Cameron, Oral Tradition, 2011
29. Duncan Cameron, Oral Tradition, 2011.
30. Alan MacNaughton, Oral Tradition, 2011.
31. John Dye, Oral Tradition, 2011.
32. Presbytery of Mull Minutes, 1744. Argyll County Council Archives.
33. Alasdair Cameron, 'North Argyll', School of Scottish Studies Archive, University of Edinburgh. PN
34. RMS Vol. X, 1105 (1667).
35. RMS Vol. VII, 272 (1610).
36. NLS Ms 475A. 'Notes from Moidart' by Father Charles Macdonald.
37. Maclean-Bristol, N (Ed) (1998). *Inhabitants of the Inner Isles, Morvern and Ardnamurchan, 1716.* The Scottish Record Society: New Series, Vol. 21. Edinburgh.
38. MacMillan, Somerled (1971), *Bygone Lochaber; Historical and Traditional.* K&R Davidson, Glasgow.
39. NRS. GD241/165. Act to vest the entailed Estate of Ardnamurchan and Sunart, 1784.
40. Tobermory Court Records, 1856, Ref. 23; 1859, Ref. 7. Argyll County Council Archives, Lochgilphead.
41. Symington Grieve (1910). 'Griddle Ine or Een, otherwise known as Griadal Fhinn' in *Transactions of the Edinburgh Field Naturalists and Microscopical Society V. 6,* 1910-11'.
42. Baptismal Records for the Ardnamurchan area, 1800-1825.
43. Tobermory Court Records; Notices of Removals, 1864-70. Argyll County Council Archives, Lochgilhead.

44. NLS. AF49/6. Schedule as to Arrears due by Tenants on Estates of Ardnamurchan and Sunart, with a List of Cottars, 1851.
45. NRS. SC59/2/11, Sheriff Court Processes (Tobermory) 1852.
46. Duncan Cameron. School of Scottish Studies Archive, University of Edinburgh. PN 1966.118.
47. Lethbridge, T.C. (1925). 'Battle Site in Gorten Bay, Kentra, Ardnamurchan' in *Proceedings of the Society of Antiquaries of Scotland* Vol. 59, 1924/5 pp. 105-8.

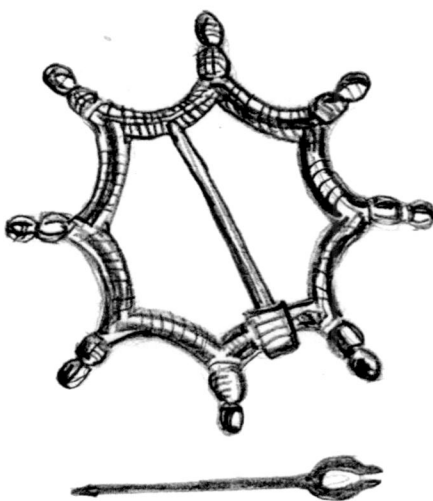

Copper alloy brooch from Cùl na Croise, Gorteneorn (actual size).
Drawn by Marilyn Gascoigne (after Lethbridge, 1925)

INDEX OF ACHARACLE COMMUNITY COUNCIL AREA
PLACE-NAMES AS RECEIVED
TOWNSHIP NAMES IN BOLD

INDEX OF GAELIC NAMES IN THE ACHARACLE COMMUNITY COUNCIL AREA

ASTERISK DENOTES HYPOTHETICAL FORMS
TOWNSHIP NAMES IN BOLD

54

Old jetty at Acharacle, destroyed in May, 2010

THE LOST PLACE-NAMES OF

ARDNAMURCHAN

SET IN A HISTORICAL, ARCHAEOLOGICAL

AND CULTURAL CONTEXT

ARDNAMURCHAN TOWNSHIPS

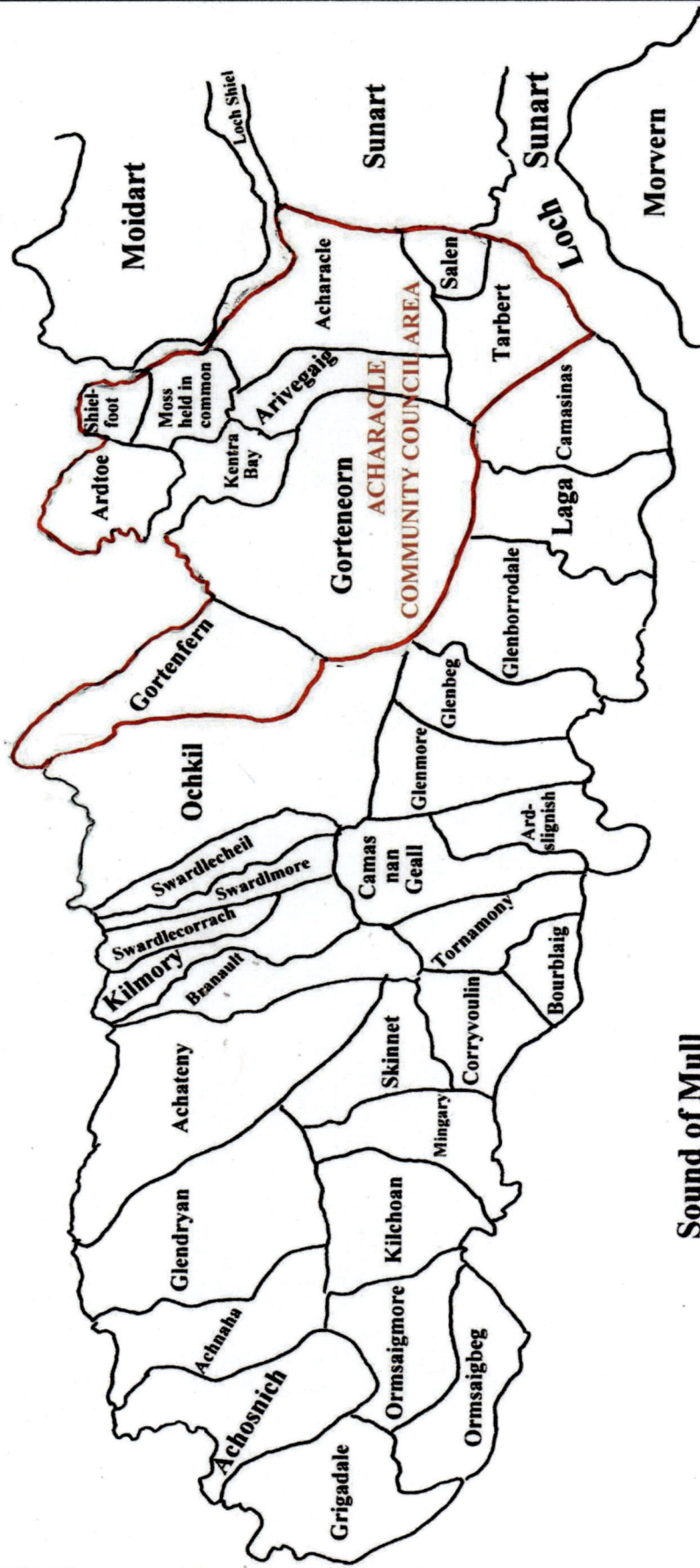

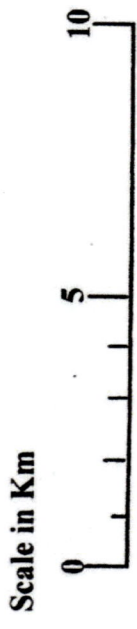

N

Moidart

Loch Shiel

Sunart

Loch Sunart

Morvern

Salen

Acharacle

Tarbert

Shiel-foot

Moss held in common

Arivegaig

Camasinas

Ardtoe

Kentra Bay

ACHARACLE COMMUNITY COUNCIL AREA

Gorteneorn

Laga

Gortenfern

Glenborrodale

Ochkil

Glenbeg

Glenmore

Swardlecheil

Swardlmore

Camas nan Geall

Ard-slignish

Swardlecorrach

Kilmory

Branault

Tornamony

Bourblaig

Achateny

Skinnet

Corryvoulin

Glendryan

Mingary

Kilchoan

Sound of Mull

Achnaha

Ormsaigmore

Achosnich

Ormsaigbeg

Grigadale

Scale in Km

0   5   10

ARDNAMURCHAN TOWNSHIPS

ACHATENY AND FASCADALE : Achadh Teine agus Faisgeadail

1541 and 1610, Faskistell, Auchatere (55, 36); 1618, Phesodill, Fastridill, Achaterie (8); 1667, Auchate and Tastkistell (5); 1686, Auchatey (10); 1723. Auchateny & Fasgaddell (3); 1723, Achateny and Faskadale (4); 1734, Achaten (34); 1737, Achtenny and Ffascadale (9); 1767, Fascado and Achateny (29); 1784, Auchatennie, Fascadell (45); 1807, Auchateny (7); 1828, Acheteny (6); 2007, Achateny, Fascadale (31).

Fascadale Bay

Valuation: Achateny was valued as a 3 merk land and Faskadale a 20/- land in 1541 and 1610 but by 1723, Achateny was a 6 penny land and Fascadale a 3 penny land.

Area:

North from which (Glendrian) about 3 miles over Hills and Glens where there is tolerable pasture for Yell Cattle, lies Fascado with a Miserable Hutt on it, above which in a glen there is about 3 acres of Arable or Grass ground and North from it on the sea side is a Moss that has been drain'd, but the Drains mostly now fill'd up. the Moss but very rough pasture above which is about 3 Acre of improved Moss ground & above that all in the Brow of the Hill for about a Mile is a scrubb wood of Birch & Hazle of little value all the above is surrounded by a Dyke belonging to the Minr. whose house is pretty good built of Stone & Lime & thatched the Office Houses are dry stone & good Repair. In the Neighbourhood of the House & about it is about 60 acres of Arable & grass ground on some of which are pityfull enclosures the soil here is very good particularly about the Ministers house where there is extreme fine grass. This is call'd amongst the best ffarms in Ardnamurchan (Anon,1767).

In 1807 the 2790.18 acres consisted of 82.34 arable, 175.09 cultivated with the spade, 1149.72 fine hill pasture and 1383.03 acres of moor & pasture.

There is on this farm some of the best arable Land on the Estate, and about 30 or 40 acres of which are well worthy of being inclosed. The West part as a Sheep Walk is high and bare. The Face of the Hill is very good Sheep Walk, and the whole is a complete Farm of the kind, and would keep Cheviot Sheep. (Alex Lowe, 1807).

Tenants: 1541, Johne McAlister McKane in Faskistell and Rore McAlister McKane in Auchatere (55); 1618, Donald Glas McEane Vane in Phesodill; Gilleane McIloure and Johnne McKinla (8); 1686, Neill McColduy VcIlory (10); 1716, Donald McIlvra, John McIlvra, John Fergusson, Ewn McKenrig, Donald McKenzie, Kenneth McKenzie, Donald McIlvoyle, John McEan VcConachie Vaidd, son to John McConachie Vaidd, alias Mcpherson, Duncan McIlvra, Duncan Mcpherson, John McInnish, a herd, Donald McCorquodill and Alexander McLachlan (37); 1737, Duncan Stewart, who also held Camas na gaul and Culoan, Ardslignish, Glenmore and Glenbeg, (9); 1807, Donald McLean (7); 1827-1834, James Thorburn (6). 1852, 6 tenants were evicted for non-payment of rent (53).

Population: 1723, Fascadale had 2 families, with 3 men, 3 women and 3 children, total 9. Achateny had 10 families, with 17 men, 21 women and 17 children, total 55.

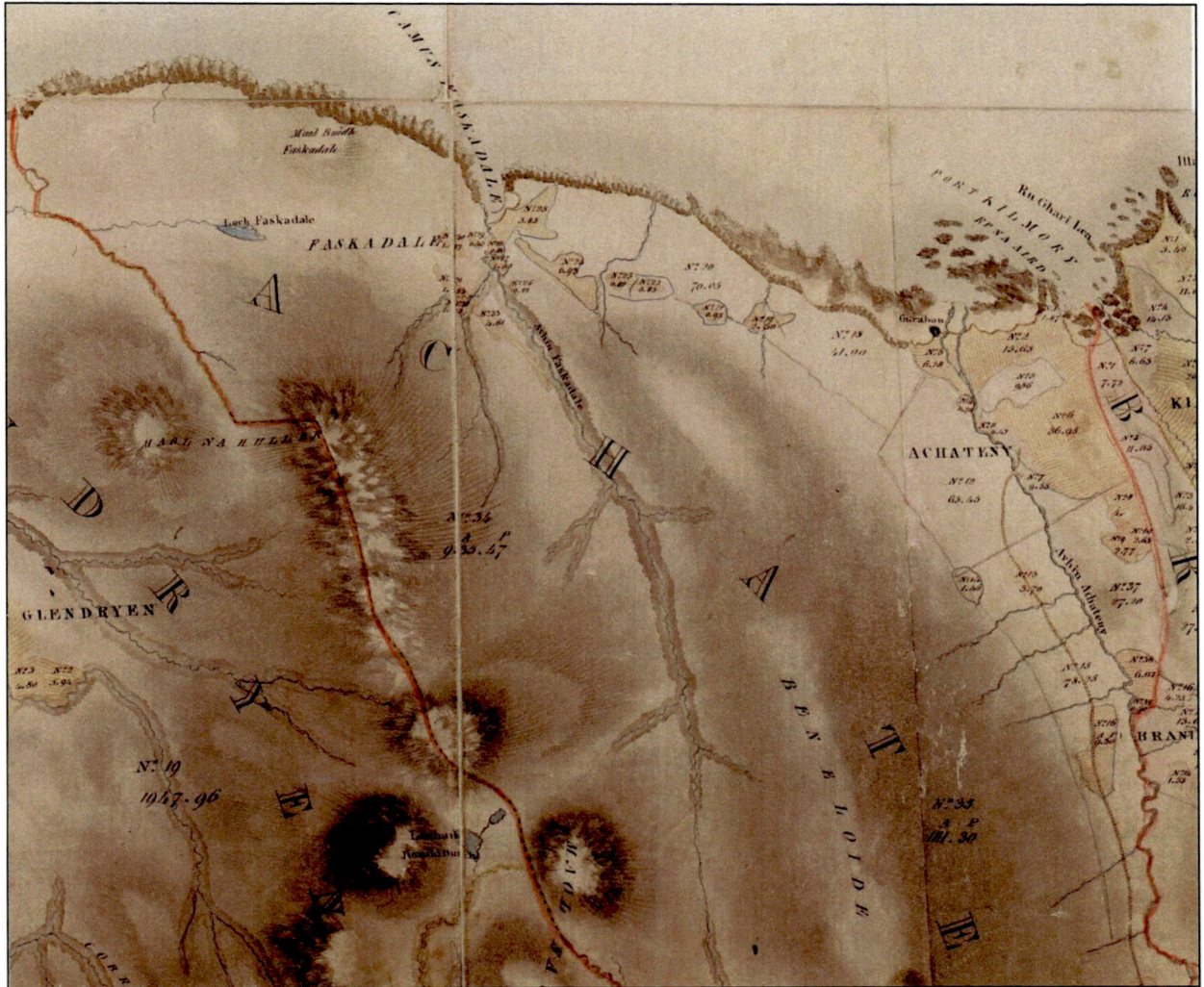

Achateny (detail)
National Records of Scotland, RHP72/1-8, Plan of Ardnamurchan and Sunart, Argyll, 1806.

Achateny fank looking towards *Dùn Mòr* on Swardale

Achateny Water looking SSW

Settlements:

NM 5220 7002. Achateny. NM 5004 7075. Fascadale.
NM 5188 7014. NM 5185 7023.
NM 5220 7001. Farmhouse, constructed 1816. NM 5195 7036.
NM 5270 6784. Braehouse. NM 5222 7069. Acarsaid Croft.
Achateny Mill; location not known, but probably at NM 5193 7020.

Shielings: Not recorded.

Placenames:

1. **Achateny (31). NM 5220 7002.** ?G. *Ach' an Teinidh* or *Achadh Teine* 'the field of the fire'.

2. **Fascadale (31). NM 5004 7075.** ?G. *Faisgeadal* 'the valley of the ship'. ?ON *aska* 'a ship' and *dalr* 'valley'. Possibly with G. *fasgadh* 'shelter; penfolding of cattle'.

3. **Camus Faskadale (1). NM 4990 7078.** G. *Camas Faisgeadail, Bàgh Faisgeadail* 'Faskadale bay'.

4. **Maol Buidh Faskadale (1). NM 4911 7110.** G. *Maol Bhuidhe Faisgeadail* 'the rounded yellow hill of Fascadale'. Named *Meall Buidhe Mòr* 'the big round yellow hill' on OS maps.

5. **Loch Faskadale (1). NM 4895 7056.** G. *Loch Faisgeadail.* Named *Lochan Dubh* 'the black lochan' on OS maps.

6. **Avhin Faskadale (1). NM 5024 7000.** G. *Abhainn Faisgeadail* 'the River Faskadale'. Named *Allt Fascadale* 'Faskadale burn' on OS maps.

7. **Ben E Loide (1). NM 5147 6894.** G. *Beinn Mhic Leòid* 'MacLeod's hill'. Named *Cathair Mhic Dhiarmaid* 'MacDermot's seat' on OS maps.

8. **Abhin Achateny (1).** *NM 5258 6935.* G. *Abhainn Ach' an Teinidh* 'River Achateny' or 'Achateny Water', and named Achateny Water on OS maps.

9. **Braehouse (2). NM 5270 6784.** G. *?Bràigh an Abhsaidh* 'the head of An t-Abhsadh'.

10. **An t'Abhsadh (2). NM 5287 6780.** G. *An t-Abhsadh.* Henderson describes this as 'A turbulent little stream called *An t-Abhsadh* which means a sudden deviation or the act of going off at a tangent'. Dwelly gives *amhasach,* (adj) 'wild'.

11. **Ru na Aird (1).** NM 5219 7090. G. *Rubha na h-Àirde* 'the headland of the height'. *Àird* gen. sing. *àrd* 'high', 'lofty'.

12. **Parcean Achateny (47).** NM 5245 7000. G. *Paircean Achateine* 'Achateny Park(s)'. Often referred to as G. *Dail Mhòr* 'the big field'.

13. **Parc na h-Ath Chuil (47).** NM 5236 7020. G. *Parce na h-Ath Chaoil* 'the field of the back (or narrow) kiln'. Sometimes referred to as G. *Pairce na Cloiche* 'the stony field'.

14. **Bac Mòr (47).** NM 5234 7040. G. *Bac Mòr* 'the big bank'.

15. **Allt an Teanga (47).** NM 5260 7020. G. *Allt an Teanga* 'the burn of the tongue'.

16. **Port Nic Choinach (47).** NM5218 7012. G. *Port Nic Choinich* 'the boat-landing of Kenneth's daughter'. The headland here is *Rubha na h-Acairseid* 'the point of the anchorage'. The anchorage is surrounded by rocks, with a restricted access, and not particularly secure. This boat-landing is in a narrow defile between rocks, but has a small quay alongside to facilitate loading, and the adjoining rocky outcrops are daubed with pitch or tar.

17. **Sgeir na Ban (47).** NM 5225 7095. G. *Sgeir nam Ban* 'the skerry of the women', with *nam ban,* genative plural of *bean* 'woman'. Genative singular would be *na mna.*

Boat noosts in Port na Banntraich with rubble from the jetty on the right

18. **Port na Boundary (47).** NM 5230 7075. G. *Port na Banntraich* 'the boat-landing of the widow'. This *port* is only accessible at high water, but there is a very ruinous quay alongside and a large double boat-noost, the only noosts recorded in Ardnamurchan, suggesting a site of considerable antiquity.

19. **Carnleadh (11).** NM ? G. *An Càrn Liath* 'the grey cairn'. The road from Achateny Bridge to Carnleadh was repaired in 1813. The name *Càrn Liath* is commonly used for Neolithic or Bronze Age burial cairns, but there is no record of one here.

20. **Guraban (1).** NM 5180 7047. G. *An Gurraban* 'the monolith'. A tall slender column of rock often found on the foreshore. The one at NM 5226 7065, near Port Kilmory, also on Achateny, fits the description better.

Rubha an Àird and *Meal Buidhe Mòr*, Fascadale from *Rubha na h-Acairseid*

Port Kilmory from the byre (?) on *Rubha na h-Acairseid*

Port Bàn, Gurraban and the march dyke between Faskadale and Achateny

Faskadale from the north

ACHNAHA : Achadh na h-Àtha : the field of the kiln

1541, Auchynhaw (55); 1610, Awchinhaw (36); 1618, Auchinha (8); 1667, Auchinhaw (or haugh) (5); 1722 Achanaha (20); 1723 Auchaha (3); 1734, Achanahaa (45); 1737, Achanaha (9); 1784, Achnaha (45); 1806, Auchnaha (7); 1828, Achnaha (6); 2007. Achnaha (31).

Achnaha from the NW with *Creag an Airgid* beyond

Valuation: Valued as a 2½ merk land in 1610, and a 5 penny land in 1723.

Area: The 1166.48 acres were made up of 90.98 arable, 21.75 cultivated with the spade, and 1055.75 moor & pasture.

Tenants: 1541, Donald McAlister McKane (55); 1618, Angus McConeill VcAngus and Angus McConeill Oig (16); 1686, Donald McInven (10); 1691, Donald McGillies and his wife Margaret NcCarmig (38); 1693, Donald and Ketherine Cameron and their son Dugald (38); 1716, Alexander McAllen VcIllespick, John McDonald, son to Alexander McDonald, Alexander McIllespie Vc Ean yearn, Archibald McDonald his brother, John McEwn VcEan roy, Duncan McEachin, Donald roy McKenrig (37); 1737, John McColl, (⅜ of the 5 penny lands), Donald McKenzie (¼), Archibald McDonald (¼) and John Stewart (⅛); 1739, Angus and Archibald McDonald (13); 1807, Donald McDonald and 6 others; 1828, let on a year to year basis to 9 tenants, and ten by 1847 (46).

Population: 1723, 8 families, with 9 men, 10 women and 14 children, total 33.

Settlements; NM 4634 6830. Achnaha. Twenty structures shown in 1897.
NM 4525 6984. Plocaig. A compact crofting clachan.

Shielings: NM 4601 6999 (two structures).

Placenames:

1. **Achnaha (31). NM 4634 6830.** G. *Achaidh na h-Àtha* 'the field of the kiln'. *Àth,* gen. *àtha* 'a kiln'.

The fortified *Rubha an Dùin Bhàin* from the SE

67

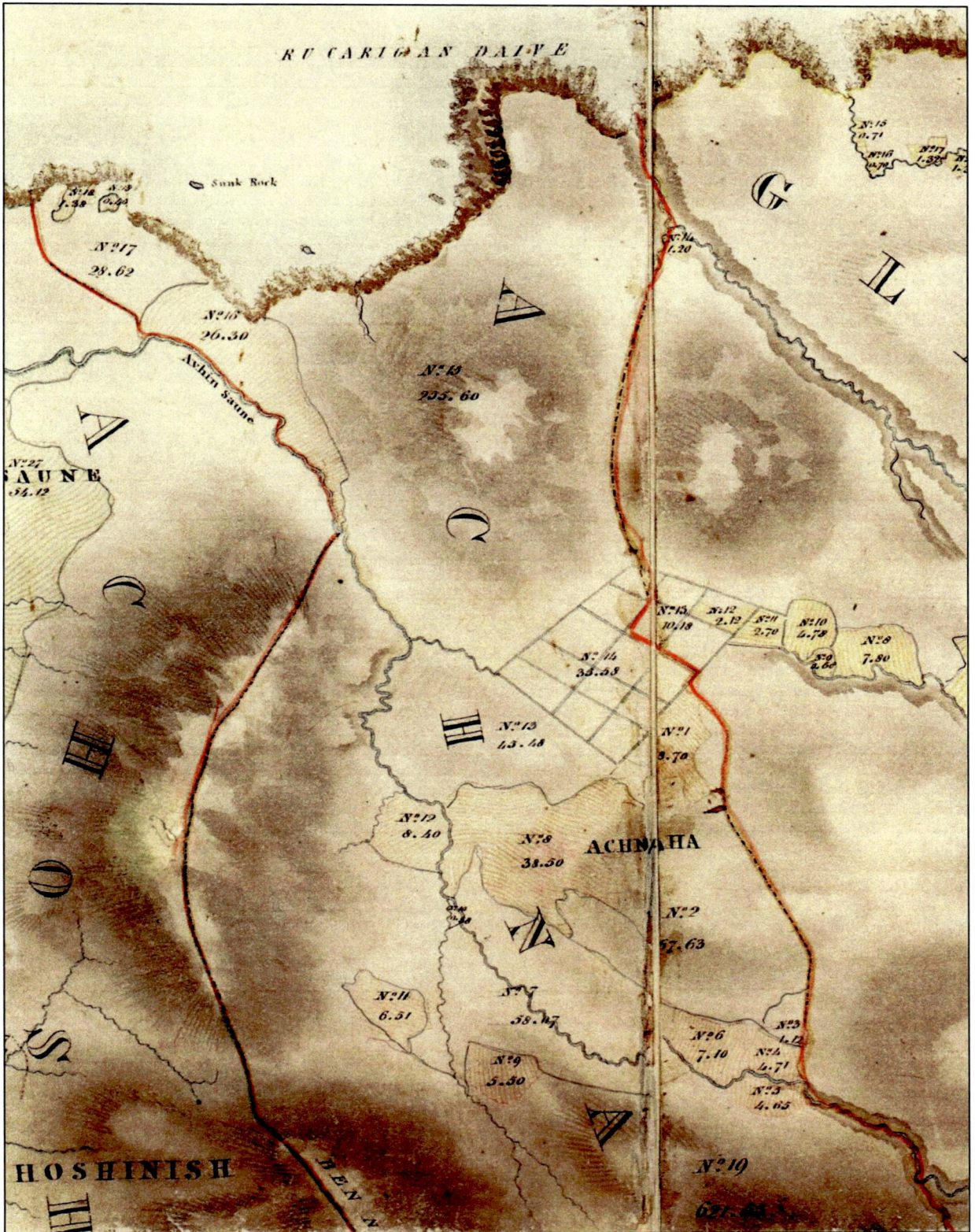

Achnaha (detail)

National Records of Scotland, RHP72/1-8, Plan of Ardnamurchan and Sunart, Argyll, 1806.

Rubha Carraig an Daimh or *Rubha Carrach*

2. **Ru Carig an Daive (1). NM 4606 7077.** G. *Rubha Carraig an Daimh* 'the headland of the stag's rock'. Named *Rubha Carrach* on OS maps. *Carraig* f 'a sea rock; fishing rock'. *Damh*, gen. *daimh* m 'a stag'.

Port Plocaig

3. **Port Plocaig (30). NM 4530 7000.** G. P*ort Plocaig* 'Plocaig port'. *Ploc* (gen. *pluic*) m 'a clod, a lump of anything'. *Plocach* (adj.) 'abounding in clods'. *Plocag* (gen. *plocaig*) f 'a corpulent little woman'. Possibly with the ON generic term *aig* 'bay'.

4. **Bo ha na craig (39). NM 4570 7084.** G. *Bodha na Creag* 'the sunk rock of the crag'.

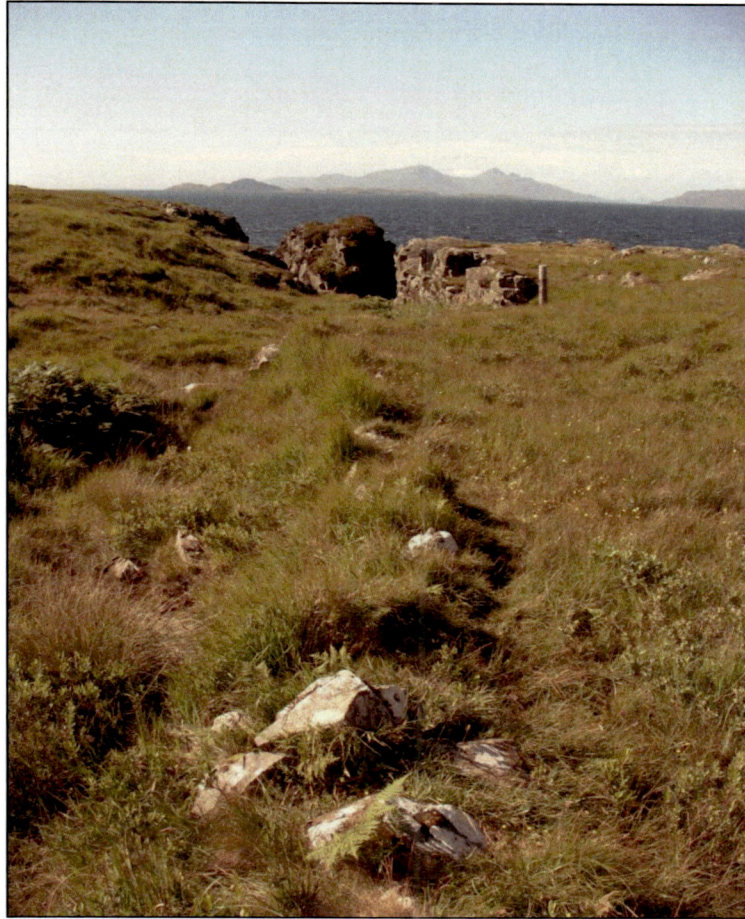

Glac na Chriche

5. **Glac Chriche (30). NM 4458 7043.** G. *Glac na Chriche* 'the hollow valley of the march'. The stone and turf march dyke between Achosnich and Achnaha (or latterly, between Sanna and Plocaig), runs through a narrow valley, and changes to a dry-stone dyke before meeting the sea at a *geo* to form a 'proper kenmarie'.

Plocaig from the SE (left) and from the NW (right)

ACHOSNICH : Achadh Osnaiche : the field of sighing

1541, Auchaquhossin; Udyne (55); 1610, Auchanquhossin, Udny, Udnye (36); 1667, Auchnaquhislick, Udyne (5); 1722 Achachosnich, Saine, Edin & Garradih (20); 1723 Auchoshnich, Eden & Sana (3); 1723, Achahosnich (4); 1737, Achachoisnich, (9); 1767, Ahosnik (Anon); 1784, Achnachosaich (45); 1807, Achosinish (7); 1828, Achosnich (6); 2007, Achosnich, Sanna (31).

Achosnich

Valuation: In 1541 and 1610, Achosnich and Edine were both valued as 2 merk lands, but by 1723 they were combined as a 9 penny land. Edine was valued as a 3 penny land in 1784, and Achosnich as a 5 penny land (36).

Area: *To the North (of Grigadale) all the way to Ahosnik which is about 2 miles over steep rugged hills & Moss in the Hollows which there (is) no chance for improving of or even planting Round this last mentioned Town lie about 8 or 10 acres of arable or grass ground mostly surrounded with a Ditch & Dyke above which is a very fine Glen for pasture, & below it about ½ a mile at the Sea Side is about 5 acre of improv'd Moss ground, and north from that about 2 miles along the Coast is a very fine Shieling of the same Town containing about 30 acre of fine Natural grass ground surrounded with a brow or Brae about ½ Mile high on the top of which is a Dyke which incloses the whole. This might be made a very pretty place & fine openings for Fish boats all along the Coast (Anon 1767).*

In 1807 the 1351.19 acres were made up of 101.00 arable, 36.10 cultivated with the spade and 1214.09 moor & pasture.

This is a very good Farm of the kind. Well shelter'd from all points, and suited to a mixture of black cattle and sheep. The low lands of Saun are very sandy; part of them is blown away and still a greater part in danger. Something to prevent this should be done without delay. Some Stake and Rice may do as has been happily tried for the same purpose at Ardtoe. This part of the farm should not be allowed to be brought into tillage. Alex. Lowe, 1806.

Tenants: 1541, Angus McAlister McKane (55); 1698, Dugald McIlphatrick, his wife Beacy NcInnish and their children Angus, Donald, John and Patrick (38); 1716, Donald McColl, Alexander McDonald younger, Donald McIlvoyle, residenter, Donald McInnish VcAlester Vullich, residenter, vagabond, John McDonald, Ronald McDonald his son, Donald McVarrish, Alexander McDonald elder (37); 1734, Archibald MacDonell, tacksman (12); 1737, James Stewart, John Stewart and Allan Stewart each possessed ⅓ (9); 1744, Ann McDonald alias McAlister van (35); 1734, Archibald McDonald, Tacksman (12); 1767, Capt. John Campbell (29); 1806, Capt. John Campbell (7); 1825-1843, let to 6 tenants (6); with eight by 1847 (46); 1852, 5 tenants were evicted for non-payment of rent (53). 1864, 1 crofter was cleared in 1864, and 3 others in 1868 (42).

Population: In 1723, 6 families with 9 men, 14 women and 11 children, total 34.

Settlements: NM 4446 6722. Achosnich. A recent crofting township.
NM 4537 6642. Aodainn. Now named Sonachan.
NM 4370 6820. Portuairk. An extensive crofting township.
NM 4471 6930. Sanna. A former crofting township.
NM 4271 6817. House, barn and byre.
Garradih, Location unknown, but possibly the name of the settlement listed above.

Achosnich (detail)   National Records of Scotland, RHP72/1-8, Plan of Ardnamurchan and Sunart, Argyll, 1806.

Shielings: Not known.

Placenames:

   1. **Achosnich (31). NM 4446 6722.** G. *Achadh Osnaiche* 'the field of sighing'. *Osnaich* (gen. *osnaiche* f) 'sighing'.

   2. **Ru Saune (1). NM 4378 7721.** Named Sanna Point on OS maps, which is the English translation. ? G. *Rubha Shanna* 'the point of Sanna', with possibly a loan-name from Old Norse *Sand-à* 'sand river'.

   3. **Avhin Saune (1). NM 4580 6980.** ?G. *Abhainn Sanna* 'the river of Sanna'. Named *Allt Sanna* 'Sanna Burn' on OS Maps.

   4. **Saune (1). NM 4450 6930.**? G. *Sanna.* A Gaelicisation of the Old Norse *Sand-à* 'Sand River'.

5. **Skeir McCormaig (1)**. **NM 4360 7016**. G. *Sgeir MhicCormaig* 'MacCormack's skerry'. Named *Sgeir Ghobhlach* 'the forked skerry' on OS maps. *Gobhlach*, adj. 'forked'.

6. **Lamrig More (1)**. **NM 4380 6986**. G. *An Laimrig Mhòr* 'the big landing place'. *Laimrig* f. 'a quay, landing place'.

7. **Island Saune (1)**. **NM 4370 6976**. ?G. *Eilean Sanna* 'Sanna Island' as on OS maps.

8. **Chorra skeir (1)**. **NM 4336 6858**. G. *An Corra Sgeir* 'Heron skerry' or G. *An Corra Sgeir* 'the pointed /projecting skerry'. Named **Sgeir Horsegate** 'the skerry of heron skerry' on OS maps. *Corra* f. 'heron, sheep'.

9. **Ru Ard Grinan (1)**. **NM 4390 6943**. G. *Rubha Àird a' Ghrianain* 'the point of the high sunny hilloch'.

10. **Port na Tuine (1)**. **NM 4444 6864**. G. *Port nan Tonn* 'the port of the waves'. The Estate Plan gives 'the port of the waves' as an alternative to the Gaelic. *Tonn* pl. *tuinn* 'waves'.

11. **Bun na Haune (1)**. **NM 4416 6824**. G. *Bun na h-Aibhne* 'the foot or mouth of the river'. Now probably named *Inbhir Allt na Luachair* 'the mouth of the burn of the common rush' on OS maps. Possibly *haune , haunn* from Old Norse *hafn* 'harbour; haven'. It is unlikely to be *Bun a' Shaune* 'the mouth of the Sand river', which is 1.5 km to the north.

12. **Edine (3), Udyne (55)**. **NM 4537 6642**. G *An Aodann* 'the face (of the hill') centered on the Sonachan Hotel area.

13. **Garradih (20)**. Possibly at **NM 4271 6817**. G. *An Gàrradh* or *An Geàrraidh* 'the garden, field wall or stony dyke'. *Garaidh* m. 'a den'. *gàrradh* (gen. *garaidh*, pl. *gàrraidhean*) m 'garden, field wall, stone dyke'.

14. **Traig Bheag (30)**. **NM 4435 6930 (?)**. G. *An Tràigh Bheag* 'the small beach'.

15. **Lon Bheinn na Curra (30)**. G. *Lòn Bheinn nan Corra* 'the moss of heron mountain'. A peat moss on the Sanna Crofting Township.

16. **Beinn na Curra (30)**. **NM?** G. *Beinn nan Corra* 'the mountain of the heron'.

17. **Toll nan Conn (30)**. **NM 4466 6921**. G. *Toll nan Con* 'the hole of the dogs'. A place in the burn below the Maclean's croft where they drowned unwanted pups. *Cù* (gen. pl. *con*) m 'dog'.

18. **Glac an Fhaing (30)**. **NM?** G. *Glac an Fhaing* 'the hollow of the fank'. Iain Maclean wrote 'Mr. F arrived to caravan. I had to lend him two planks to get across the stream at *Glac an Fhaing* with his car'. *Glac* f. 'valley, hollow'. *fang* (gen. *fainge*) f. 'fank'. G. *Glac na Fainge*, unless the specific is m. here: *Glac an Fhaing*.

Croft houses and corrugated iron church or hall, Sanna

19. **Church or Hall (30). NM 4475 6894.** 'Services in our little Hall or church, actually a corrugated iron hut with a few rows of seats, were held on alternate Sundays by the Church of Scotland (otherwise known as the Parish Church or the Established Church) and the Free Church, a Calvinistic breakaway group'.

20. **Cnoc Breac (30). NM 4547 6975?** G. *An Cnoc Breac* 'the speckled round hill'. Iain Maclean wrote 'managed across river at *Cnoc Breac* peats'.

21. **Beinn Dubh (30). NM ?** G. *A' Bheinn Dubh* 'the black mountain'.

22. **Druim (30). NM 4445 6936.** G. *An Druim* 'the ridge'. This was a plateau of consolidated dunes, with evidence of past occupation, but it has all but vanished due to the action of rabbits and tourists. Pieces of worked flints and mediaeval pottery occasionally turn up, and the Kelvingrove Museum has pieces of a bronze horse harness that was found near the burnside in the 1960s.

23. **Gorten (30). NM /?** G. *An Goirtean* 'the small field' or 'the enclosed corn-land'.

24. **Dalmore (30). NM 4500 6940.** G. *An Dail Mhòr* 'the big field'. This was the biggest field on the Maclean family's croft.

25. **Bealach Faotidh (30). NM 4503 6834.** ?G *Bealach Futaidh* 'the windy pass'. *Futadh (*genetive *futaidh*) m. 'blustering, windy'. 'A pass through the range of hills that separates the hamlets of Achnaha and Sanna. It makes a fine short cut between the two places but only for such natives as have glossy black wings and a harsh voice'.

26. **Bo Kora Ban (40). NNM 4235 6835.** G. *Bodha na Corra Bhàn* 'the sunk rock of the white heron'.

27. **Portnarick Crofts. (32).** Named on the Valuation Roll of 1873. This looks like a mistake for Portuairk.

28. **Port na Gruaigean (21). NM 4350 6820** (township) and **NM 4392 6811** (port). G.*Port na Gruaigean* 'the port of the bladderlocks or henware', an edible seaweed. This is said to be an earlier name for **Portuairk,** possibly G. *Port nan t-Sòbhrach* 'the port of the primroses'.

29. **Lochan Cuilc (30). NM 4470 6902.** G. *Lochan na Cuilc* 'the pool of the reeds'. A smallpox epidemic swept through the area in 1830, and Neil Macdonald, a Portuairk man, who appeared to have built up a resistance to the infection, undertook to deal with those in Sanna who had succumbed, and this is where he stripped and bathed before going home. The 'reed' was *Phragmites commune,* valued as a thatching material in the past.

A croft house north of the *Allt Sanna*

ARDSLIGNISH : Ard Slignis, Ard Slige-innis or Ard Sliognis : height or point of the shelly place

1541, Ardslegingish (55); 1610, Ardslegingis, (36); 1618, Ardsleignes, Ardstiginshe (8); 1667, Ardsliggensh (5); 1722 Ardslignisk (20); 1723; Ardslignish (4); 1723, Ardsleegnish, (9); 1723, Ardslignish (3); 1784, Ardslignish (45); 1806, Ardslignish (7); 1826, Ardslignish (6); 2007, Ardslignish (31).

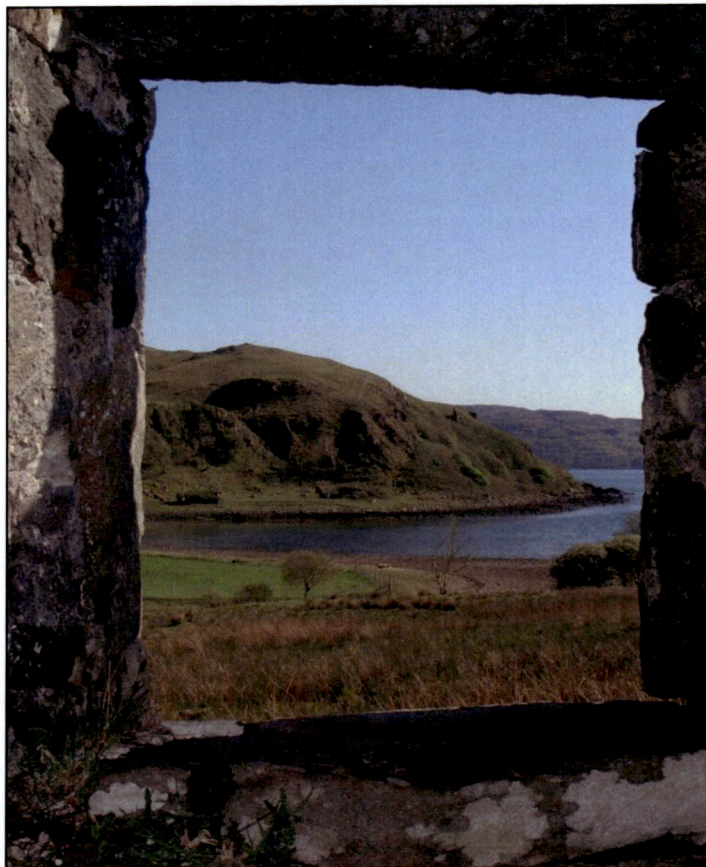

Ardslignish from Tornamony

Valuation: Valued as a 2½ merk land in 1610, and as a 5 penny land in 1723.

Area: In 1806 it contained 992.67 acres made up of 83.53 arable, 29.79 cultivated with the spade, 853.15 moor and pasture and 26.20 acres of wood, and was let with Camas nan Geall at this time.

Betwixt the (Camas nan Geall) Houses & the sea after ascending a Steep Brae lies a place calld Ardslignish surrounded on all sides by the sea except the above mentioned Brae it is just now in grass which is good & the soil appears to be of a very good kind, & contains about 20 acres no houses standing on it but two barns. The shore exceptionally (?) high & steep on all sides (Anon, 1767).

Population: 1723, 5 families, with 10 men, 8 women and 14 children, total 32.

Tenants: 1541, Rore McRanald McKane (55); 1618, Donald McAngus VcEane (8); 1690, Eun Cameron (brother to John of Glendessary) and Mary McKain his wife, with their children Allan, Margaret, John, Donald, Eune, Alexander and Archibald (38); 1696, Eouin McEouin VcMalcolm, his wife Beaig NcIllomish and their son Allan (38); 1716, Alexander Campbell, Ewn McDiarmid, a vagabond, John McCallum, Duncan McCalllum, a young lad, Robert Mcpharllan, John McIlrioch, John McEan VcEchine, John McKendrig (37); 1737, Alexander Campbell of Ardslignish (9); 1744, Mary McLauchlan, spouse to John Simpson, gardener in Ardslignish (35); 1807, Archibald Murray (7); 1811-1829, Lachlan McDonald – together with Camasingaal (3).

Ardslignish   National Records of Scotland, RHP72/1-8, Plan of Ardnamurchan and Sunart, Argyll, 1806.

Settlements: NM 5650 6125. Ardslignish,
  NM 5635 6101. Chapel and burial ground, dedicated to St. Ciaran.
  NM 5725 6198. Field system and four buildings.

19th century croft-house on the shore

Shielings: Not known.

Placenames:

1. **Ardslignish (31). NM 5650 6125.** G. *Ard Slige-innis* or *Ard Slignis.* Henderson (2) translates this as 'The height of the sheltered valley frequented by cattle and abounding in shellfish !', but *–nish* is probably a Gaelicisation of Old Norse *Nes* 'ness' or 'promontory'. *Àird* f. 'a high place, a height, promontory or point'. *Sligeanach* 'spotted, greenish coloured'; 'the spotted promontory'? Possibly *Slige-innis* 'shell, or hull of a vessel'. *Innis* 'island, field to graze cattle in, pasture, resting place for cattle, headland'. Two groups of skerries here are named *Sligneach Mòr* and *Sligneach Beag* 'big and small shelly place', so 'height or point of the shelly place' is the preferred translation.

2. **Illan an Deare (1). NM 5710 6118.** G. *Eilean nan Deur* 'the island of the tears'? with *deur* gen. *deòir* ' a drop, tear'. This interpretation seems improbable and it could be a crude attempt at *Eilean Tioram*, although today it is the smallest of skerries.

3. **Creag na Sgulan (1). NM 5608 6094.** G. *Creag nan Sgùlan* 'the rock of the fishing line'. *sgùlan,* gen. sing. *sgùlain* m. 'a fishing line, basket or creel'.

4. **Tobar Chaluim Chille (17). NM 5662 6158.** G. *Tobar Chaluim Chille* 'St. Columba's well', or literally, 'the well of the dove of the church'. A small rock-cut basin by the side of the road from which St. Columba is said to have taken water to baptise a local child in the 6th century, a use to which it is still put on occasion.

5. Old Burrying Ground (1). Cladh Chiarain (26).NM 5635 6101. G. *Cladh Chiarain* 'St Ciaran's burial ground'. This burial ground is associated with St. Ciaran, an Irish monk who died in Ireland on 9th September, 548. It consists of a D-shaped enclosure 37m NE/SW by 26m transversely with walls c.1.8m thick and an entrance in the NW sector. On the W side is a length of very ruinous walling c.9.5m long, orientated E/W. possibly the remains of an early church, and a few upright stones which could be grave-markers.

Ardslignish and Camas nan Geall

Ardslignish from the sheep fank

BOURBLAIG : Buaorblaig :

1541, Bordblege (47); 1610, Bordblege, Bordbledge (36); 1667, Borbledge (5); 1722, Bourblaig (20); 1723, Buarblaig (3); 1723, Bourblaige (4); 1724, Boorblack (47); 1737, Bourblack (5); 1748, Borblack (43); 1784, Buarblaige (45); 1820, Borblaig (42); 1828, Bourblaig (6); 2007, Bourblaige (31).

Bourblaige

Valuation: Valued as a 2½ merk land in 1610, and as a 5 penny land in 1723.

Area:
To the west of Tournamon lies Boarblack a very extensive Farm having betwixt it & the sea on a slop a fine arable Glen of about 10 acres and about the Houses 2 or 3 more the Houses all built with stone. The Soil of both these ...a light sandy kind. From this place to that calld McLean's Nose are excessive high rocks with good pasture on the top of them & some betwixt them & the sea (Anon, 1767).

In 1806 it contained 553.15 acres, with 31.75 arable, 62.76 cultivated with the spade and 458.64 acres of moor and pasture.

Population: 1723, 7 families, with 9 men, 11 women and 15 children, total 35.

Tenants: 1541, Alester McRanald McKane (47); 1686, John Cameron (10); 1698, Jon Cameroun, Mary, his wife and Duncan and Kathrein, their children (58); 1716, Ewn Mcphaill, Donald Mcphaill his brother, John Campbell, Ewn Campbell, John Mcphaill, John McKendrig, Donald bane Mcphaill (37); 1737, Duncan Stewart, along with Skinad, Corrivullin, Tornamony and Faskadale (9); 1807, John Stewart and nine others (8); 1828-1833, let to John McColl along with Tornamony;

Settlements: NM 5470 6216. Bourblaig.

Shielings: NM 5350 6219. Five oval dry-stone structures.
                  NM 5363 6262. Four turf ruin mounds.
                  NM 5353 6252. Two turf ruin mounds.

Placenames:

   1.   **Bourblaige (31). NM 5470 6216.** This name appears elsewhere in the West Highlands, but the meaning is not known.

2. **Ru Stron Chloinn an Lean (1). NM 5330 6150.** G. *Rubha Sròn Chloinn 'illEathainn* 'the point of Clan Maclean's headland'. *Sròn* 'nose, promontory, cape, headland, shoulder of a hill, buttress'.

3. **Stron Chloinn an Lean (1).** NM 53300 6160. G. *Sròn Chloinn 'illEathainn* 'Clan Maclean's buttress'. The OS maps record it as Maclean's Nose

Bourblaige   National Records of Scotland, RHP72/1-8, Plan of Ardnamurchan and Sunart, Argyll, 1806.

4. **Stron Mhic Illeain (2).** NM 5330 6150. ?G. *Sròn MhicGill'illEathainnn* 'Maclean's Nose'.

Bourblaige, with the fort on the knoll in the distance

The settlement with *Stallachan Dubha* beyond

Shieling below the *Stallachan Dubha* on *Sròn Mhòr*

BRANAULT : Bràigh nan Allt : the brae of the streams

1541, Branalt (55); 1610, Branalt (36); 1667, Branalt (5); 1716, Braynanalt (37); 1723, Braynanault (4); 1723, Brynanault (3); 1732, 1739, Braynanault (13,14); 1767, Brownhault (29); 1806, Branault (7); 1828, Branault (6); 2007, Branault (31).

Branault

Two croft-houses near the march with Achatenny

Valuation: Valued as a 2 merk land in 1541, and as a 5 penny land in 1723. In 1541, *ane half mark land of Branalt fest to Our Lady* (St. Mary) (55). This may have been Iona Priory (Nunnery), which belonged to the Augustinian Canonesses, and dedicated to St. Mary.

Area: In 1806 the 687.06 acres were made up of 84.82 arable, 31.11 cultivated with the spade, 45.83 of pasture and 525.30 acres of moorland.

About a mile & a half above which (Achateny) in a glen lies Brownhault in which there a good many houses & in tolerable repair being built with Stone and Feal, surrounded with about 20 acres of arable & grass ground, & above it about ½ a mile in the glen about 20 acre more of Haughs, & above them on the Brow of the Hill a large quantity of Scrubb Birch. The Glen fine for pasture & of a great extent (Anon, 1767).

Population: 1723, 8 families, with 7 men, 12 women and 12 children, total 31 (4);

Tenants: 1541, Johanni (55); 1716, John McEan VcConachie Vaidd, son to John McConachie Vaidd alias Mcpherson, Duncan McIlvra, Duncan Mcpherson, John Mcinnish, herd there, Donald McCorquodill, Alexander McLachlan (3); 1732, John McLachlan, Archibald McEachern and Duncan McLachlan (14);1734, Kenneth McKenzie, Ronald McDonald and Archibald McDonald (45); 1739, Allan Stewart (13); 1806, Donald McGilvray and 4 others, with John Stewart on the croft (7); 1827, let to 8 tenants on a year to year basis, but John Stewart and Donald McDonald to be

82

RU A FAOLIN

Illan Ad Toe

CAM

Ru Ghari Lea

PORT KILMORY

RUNA AIRD

RUNA AIRD

RU NA AIRD

SWORDLE CHORRA

raban

ACHATENY

KILMORY

DONMORE

N:12
65.45

N:2
13.65

N:3
6.28

N:5
9.56

N:6
36.95

N:7
9.35

N:1
7.72

N:7
6.65

N:6
20.45

N:6
20.6

N:3
3.81

N:3
3.81

N:1
3.60

N:2
8.75

N:3
12.97

N:20
35.72

N:2
31.30

N:9
6.80

N:10
6.15

N:11
4.45

N:12
9.74

N:13
9.74

N:8
4.

N:9
2.77

N:10
2.65

N:13
3.70

N:37
27.20

N:38
6.02

N:15
78.25

N:16
2.50

N:16
4.25

N:13
13.60

N:13
15.57

N:14
1.53

BRANAULT

N:16
27.80

N:5
7.86

N:14
7.60

N:16
9.17

N:17
12.93

N:6
6.50

N:7
3.93

N:8
8.20

N:9
10.92

N:17
5.68

N:10
9.20

N:15
5.40

N:19
A P
752.40

Avhin Achateny

DRIM SCAILE

BORDE

SWORDLE M

N:17
A P
252.90

N:33
A P

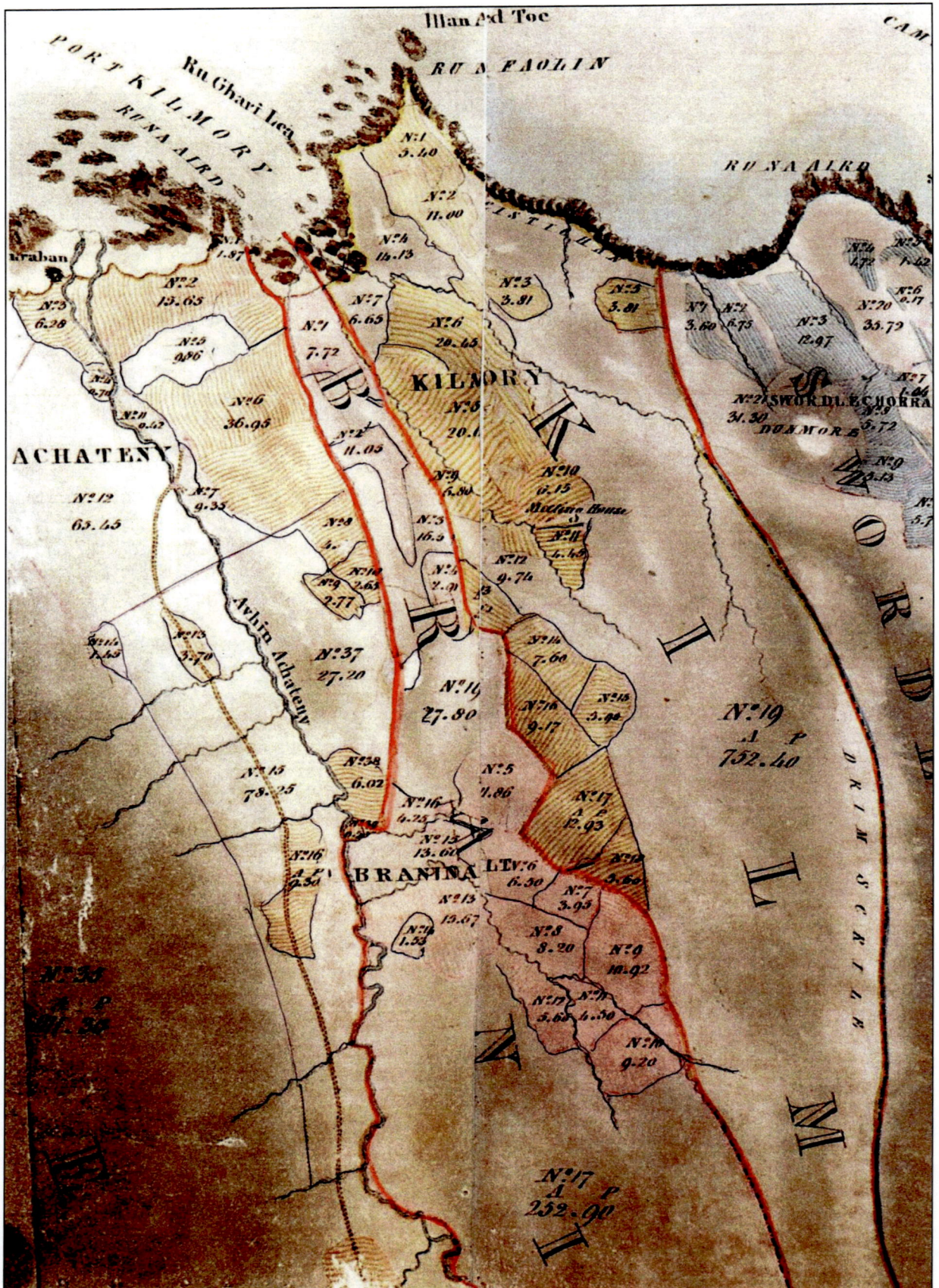

Branault and Kilmory (detail)   National Records of Scotland, RHP72/1-8, Plan of Ardnamurchan and Sunart, Argyll, 1806.

removed. *John Stewart ought to be allowed to retain his house, with a couple of cows. He is now getting infirm and really an object of Charity* (6). Let to 6 tenants in 1847 (46). 1852, 6 tenants were evicted for non-payment of rent (53).

Settlements: NM 5290 6925. Branault.

Shielings: NM 5393 6616. 'Doire Daraich'. A developed shieling with several buildings and a field system.

Placenames:

1. **Branault (31). NM 5290 6925.** G. *Bràigh nan Allt* 'the brae of the streams' or 'above the burn(s).

Guraban, a prehistoric standing stone

2. **Guraban (2). NM 5265 6949.** G. *Gurraban,* 'a tall slender column of rock' or 'standing stone'. Described as a 'standing monolith'. The RCAHMS describes it thus: 'There are two standing stones of basalt in a hollow between a pair of rock ridges 460m NW of Branault. One stone measures 2.2m in height and 1m by 0.6m at the base; the top is rounded. The other stone, 0.9m to the NW, is now only a stump measuring 0.5m in height and 0.5m by 0.4m at the base (28).

3. **Cladh Chatain (28). NM 5254 6966 ?.** G. *Cladh Chatain,* 'St. Catan's burial ground'. An ancient burial ground dedicated to St. Catan, an Irish monk who died on the 17th May, 595. If the Grid Reference is correct then the site is a very ruinous sub-rectangular enclosure whose western wall is formed by a ridge of rock. The interior is wet, and the grave slabs, which are known to have been procumbent, have been covered by peaty soil. An even smaller and more ruinous enclosure a short distance to the SSE may contain the fragmentary remains of a chapel or cell orientated E/W.

4. **Teanga na Mall (47). NM 5305 6988.** G. ?*Teanga nam Màl* 'the tongue of the place of the rents'. Could this have been the place where tenants met on rent day?

84

The small section of beach at the head of Port Kilmory belonging to Branault

Field system looking towards Eigg and Skye

The developed shieling of Doire Darraich

CAMASINAS : Camas Aonghais : Angus's bay

1541, Cammustynnes and Cammustorsay (55); 1610, Cammuscynees et Cammustorsay (36); 1722, Camisinis (3); 1722, Caminish (3); 1737, Cammisinnish (9); 1741, Camufs Inish (12); 1847, Camusines (46); 2007, Camasinas (31).

A house and field above the bay, looking NE to Ben Resipole

Valuation: In 1610, Camasinas and Camastorsa combined to form a 10/- (=1½d) land (36).

Area: In 1807 the 1351.97 acres were made up of 15.40 acres ploughable, 5.05 cultivated with the spade, 3.91 pasture, 128.81 woodland and 1198.80 acres of moor.

Tenants: 1737, Alexander Cameron together with Laga (9); 1740, Donald MacPherson, residenter (12); 1828-1847, McKenzie (6).

Population: Not known

Settlements: NM 6568 6108. Camasinas.
NM 6553 6095. Building(s).
NM 6575 6136. Sheepfank.
NM 6565 6113. Sheepfank and building(s).

Shielings: Not known

Placenames:

1. **Camasinas (31). NM 6568 6108.** G. *Camas Aonghais* 'Angus's bay'. Camasinas and Camastorsa together with Tarbert were considered part of Sunart until 1722-3 when they were exchanged in an excambion for the Lands of Letterlochshiel.

2. **Maol Tumaguind (1). NM 6605 6260.** ?.G. *Maol Tom na Gainntir* 'the great bare round hill of the thicket of the prison ?'. Named *Meal Tom a' Ghanntair* 'the hill of the prison hillock' on Ordnance Survey maps. *Tom* (gen. *tuim*)m. 'a round hillock, a thicket or bush'. *Gainntir* f. 'prison'.

3. **Ru Beul Maruadh (1). NM 6513 6025.** ?G. *Rubha Beul nam Màirneach* 'the promontory of the salmon's mouth'. *Rubha/rudha* m. 'headland, promontory; *beul* m. 'mouth; *màirneach* 'a full salmon'. The alternative name given on the Estate Plan is 'O-WOGS-ME', or as read by 'North Argyll', 'O WOE IS ME', which is more plausible. For the folk etymology, cf. next item.

4. **Rubha Bheul ma Thruaighe (18). NM 6514 6024.** G. *Rubha Bheul na Truaighe* 'the point of the mouth of misery'. *Truaighe* 'misery, woe, mischief, evil.calamity or wretchedness'.

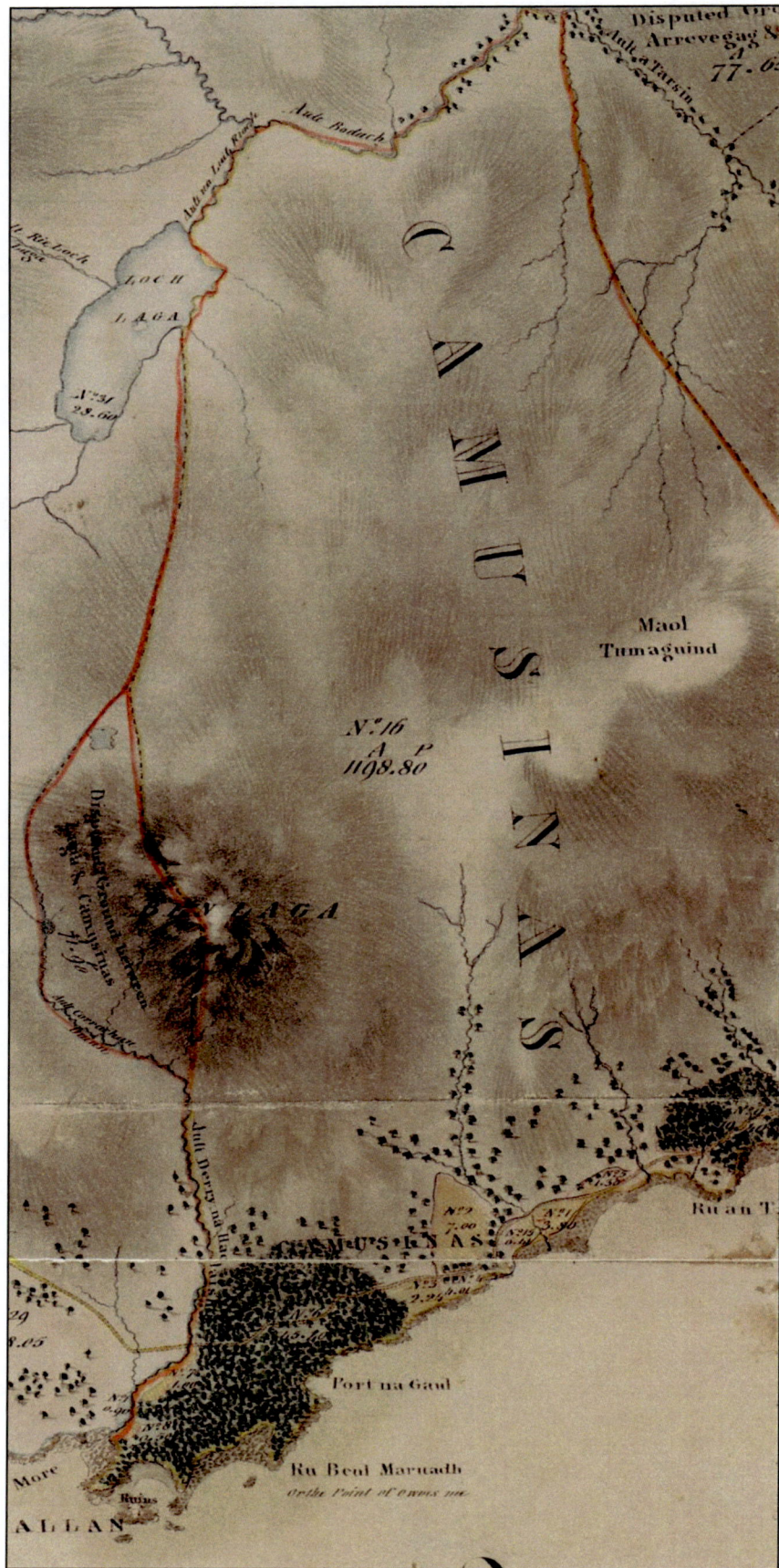

Camasinas (detail)

National Records of Scotland, RHP72/1-8, Plan of Ardnamurchan and Sunart, Argyll, 1806.

Caisteal Dùn Ghallain and *Port Mòr* on Laga

5. **Castle Dungallon (1). NM 6074 6003.** G. *Caisteal Dùn Ghallain* 'Dungallon Castle'. *Dùn a' Ghallain.* 'the fort of the hero', *gallan* 'a hero' or *Dùn na Gaillinn?* 'the fort of the storm'. *Gaillean* gen, *gaillinn* f. 'storm'. The name appears in block capitals on Bald's Estate Plan of 1807 and it is assumed that the Iron Age dùn was refortified in the Mediaeval Period, to protect the safe harbour and (?) boat repairing facilities.

6. **Beallach Slochd an Each (18). NM 6554 6216.** G. *Bealoch Sloc nan Each* or *Bealach Sloc an Eich,* 'the horse's hollow pass'.

7. **Am Breun Torr (18). NM 6522 6060.** G. *An Breun Torr* 'the stinking hill'. *Breun* (gen. *brèine*) f. 'stench, stink'; *breun* (adj.) 'stinking, filthy, turbulent'. *Tòrr* (gen. *torra*) m. 'a small hill, a mound or heap'.

8. **Port na Caul (15).** NM   G. *Port nan Càl* 'the port of the kail' or 'cabbages'.

One of nineteen recessed platforms recorded to date on Camasinas as seen from the SE (left),
and overlooking the house and field system from the NE.

Croft house above Dungallon

Camas Aonghas

CAMAS NAN GEALL: Camas nan Ceall : the bay of the churches

1541, Cammusnynggald (55); 1610, Camusnyngald, Cammusnynggald (36); 1618, Camisnagald, Tamisingill (8); 1686, Camisnogow, Camisuogeall (10); 1722, Camasingaul & Corrynaneach (3); 1723, Camisangaal (4); 1737, Camasnagaoul and Culoan (9); 1767, Camisingal (29); 1784, Camisnangarl (45); 1807, Camusingal (7), Camus an gaal (1); 1827, Camusingaal (6); 2007, Camas nan Geall (31).

Camas nan Geall

Valuation: Valued as a 2½ merk land in 1610 and a 6 penny land in 1723.

Area:
In 1806 it contained 1052.88 acres, made up of 44.44 arable, 22.94 cultivated with the spade, 940.35 moor and pasture and 45.15 acres of lochs, and was let with Ardslignish.
From Glenmore Camisingal being about 2 Miles is a very high Brow of a hill for the most part covered with Hazle & Aller Brush, fine pasture tho excessively steep; here is a large Inclosure by a Stone Dyke which goes up & in the head of the Brow & encloses the most part of it containing about 80 or 100 acres – here a Gentleman lives of the name of Campbell a half pay officer who has good houses some of them built with Stone & Lime, above the Houses is a most Excellent Glen for pasture of all sorts & of great extent & hard by them are about 5 or 6 Acres of arable Ground in the midst of which is a burial place where there are some stones with inscriptions (Anon, 1767).

Tenants: 1541, Donald McAlister McKane (55); 1618, Alexander McRonald VcEane (8); 1686, Donald and Anna Cameron, Neill McOnlea (10); 1716, Archibald McDonald, Ronald McDonald, John McKendrig, Normand McLeod, John McLea, Dushlea McLea, Archibald McIlrioch, Duncan McKendrig, Duncan McLea (37); 1737, Alexander Campbell of Ardslignish (9); 1807, Archibald Murray (29); 1811-1829, Lachlan McDonald (6).

Population: 9 families, with 9 men, 13 women and 13 children, total 35 in 1723.

Settlements: NM 5605 6195. *Camas nan Geall.*
              Corrynaneach. Not located but possibly one of those listed below..
              NM 5580 6542. 'Lochan a' Mhadaidh Riabhaich'.
              NM 5505 6545. 'Lochan na Gruagaich'.
              NM 5602 6504.
              NM 5534 6350. 'Culoan?'.
              NM 5596 6218.

Shielings: Not recorded.

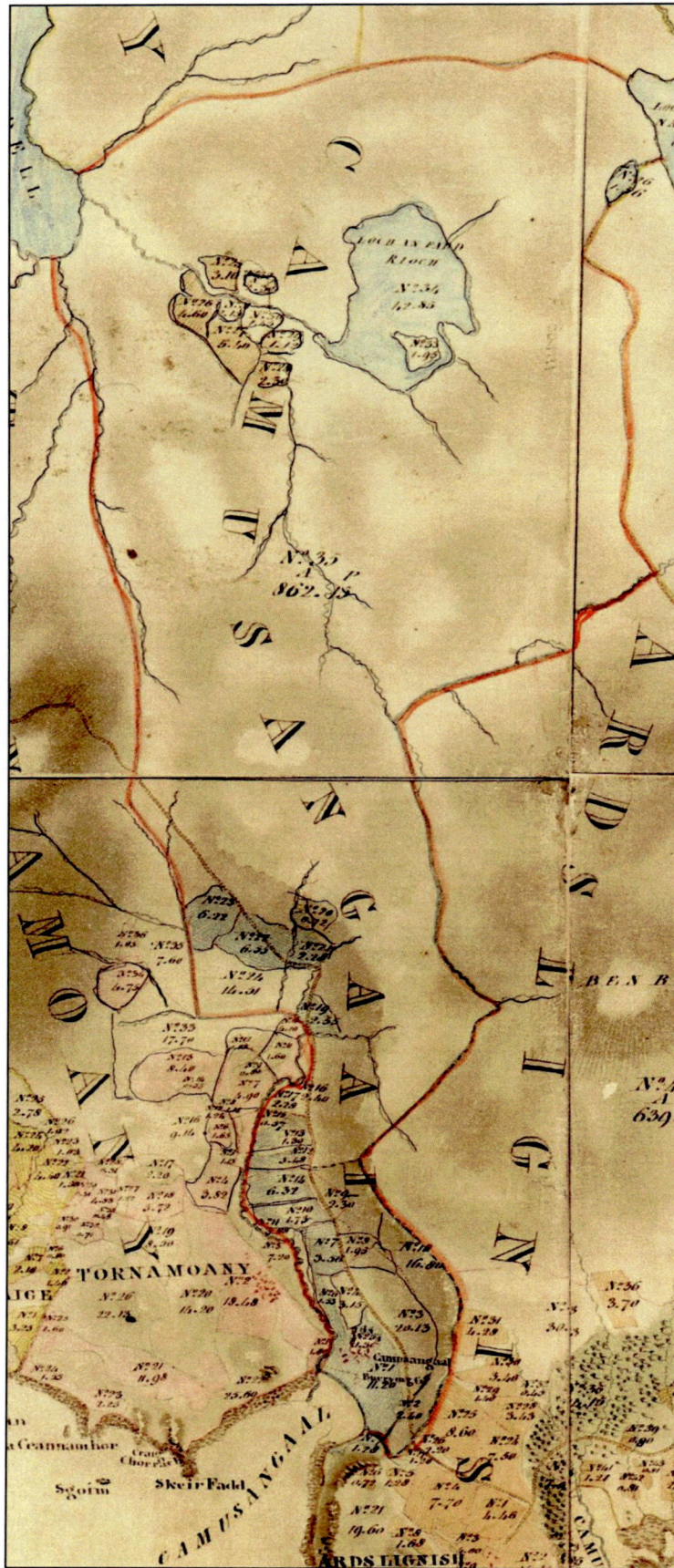

Camas nan Geal   National Records of Scotland, RHP72/1-8, Plan of Ardnamurchan and Sunart, Argyll, 1806.

House built onto the Neolithic Chambered Cairn

Croft house on Camas nan Geall

Placenames:

1. **Camus an gaal (1). NM 5605 6195.** The meaning of this name is obscure, and four possibilities have been suggested. *Camusnagaul* 'the bay of the stranger or foreigner', *Camus-na-geola* 'the 'bay of the dingy', or *Camus na Geall* 'the bay of the pledge, promise or wager' (3). Henderson (12) favours *Camus nan Cille* 'the

bay of the churches', as does Mary Cameron (49). Peadar Morgan has also suggested that this is *Camas nan Ceall,* locally *Camas nan gCeall* 'the bay of the churches'.

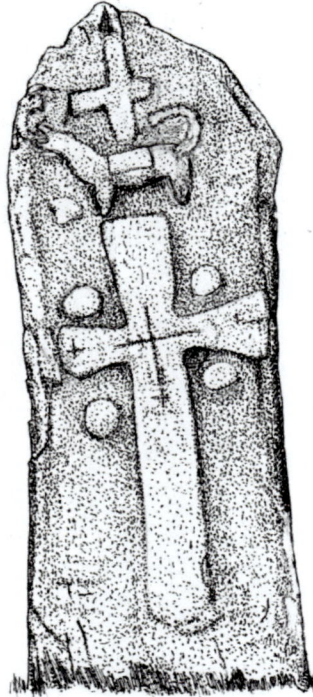

Bronze Age standing stone with Early (?) Christian motifs
outside the Campbell of Ardslignish burial ground on Camus nan Geall

2.   **Cuilowen (2). NM 5608 6195. G.** *Cùl Eòghainn* 'the nook of Ewen'. Named on the Valuation Roll of 1873/74. *Cùil* 'corner, niche, out of the way place or nook'. *Eòghann/Eòghainn* ' Ewen'. **Culorne (2).** ?G. *Cuil Eodhrain* 'the nook of Eadhran ?'.

3.   **Loch an Fadd Rioch (1)**. G ? Named *Lochan a' Mhadaidh Riabhaich* 'the lochan of the brindled dog' on OS maps, and the name on the estate plan is thought to be a phonetic rendering of this.

Camas nan Geall from Ardslignish

4. **Corrynaneach (3).** G. *Coire nan Each* 'the corry of the horses'. This settlement probably formed part of *Camas nan Geal*, but the location is unknown, although it could have been an earlier name for No. 2 **Cuilowen.**

5. **Bealach-a-bhearnas (26).** NM 5511 6410. **?**G. *Bealach a' Bheànaiche* 'the pass of the notch'.

6. **Tigh Ban (26).** NM 5605 6195. G. *An Taigh Bàn* 'the white house'. Alastair Cameron suggests that this was the name given to the original Ardslignish House. The Campbells of Ardslignish lived in a substantial house within the tree-lined enclosure at Camusnageall, as shown on General Roy's Military Survey of 1748-54. The name suggests that it may have been harled and white-washed.

A settlement on Camas nan Geall, possibly part of Culorne/Cuilowen

The Neolithic chambered cairn from the SW

CORRYVOULIN : Coire a' Mhulinn : the corrie of the mill

1541, Correwillane (55); 1610, Correwillane (36); 1667, Corriwillane (5); 1722, Correwulin (3); 1723, Corrythulin (3); 1723, Coriuolline (4); 1784, Corriewilline (45); 1806, Corrivolline (7); 1828, Corrivuline (6); 2007, Coire Mhuilinn (31).

Coire Mhuilinn looking west to Mingary

Valuation: Valued as a 2½ merk land in 1610 and a 5 penny land in 1723.

Area:

In 1806 it contained 890.19 acres, made up of 71.85 arable, 15.20 cultivated with the spade and 809.10 acres of moor and pasture. *This is a very good Sheep Farm, but also oppressed with a number of Tenants. To arrange properly, these last four Farms* (Tornamony, Bourblaig, Skinid and Corryvoulin) *should be in one and stocked with Cheviot Sheep, but probably the Rise of Rent would not be so much at the Outset as to justify the removal of so many small Tenants.*

Tenants: 1541, Donald McAlister McKane (55); 1690, Janet Cameron and her children, William, Alexander and John (38); 1716, John bane McCallan, Ewn McCallan, his son, Duncan McCarmaig, Ewn bane McKendrig, Angus McIntailor, Tavish McKendrig (37); 1739-45, Alasdair Mac Maighstir Alasdair (33); 1806, Archibald McMillan and nine others (7); 1828, John McColl (6).

Population: 1723, 7 families, with 9 men, 11 women and 9 children, total 29.

Settlements: NM 5145 6324. Corryvoulin. Nineteen structures on the estate plan of 1807.
NM 5115 6327. House and byre?

Shielings: NM 5349 6412. Eight turf structures and two with drystone walling.

Placenames:

1. **Corryvoulin (33). NM 5120 6316.** G. *Coire a' Mhuilinn* 'the corrie of the mill'. There is no archival evidence for a mill here, and neither does it appear in the archaeological record.

2. **Carn Mhuicragain (22).** In the early 13th century, Cléamainn (or Eoghan) Cléireac lived at the foot of Ben Hiant with his wife and family. His wife was said to be the most beautiful woman in Argyll, and as virtuous as she was beautiful. Her reputation as a beauty came to the attention of Muchcragan, the licentious son of the King of Lochlan, who arrived in the district to prove that her virtue could not withstand his amorous advances. Cléamainn met Muchcragan on the north side of Ben Hiant and subjected him to jeers and taunts. This was too much for the vikings and the fleet-footed Cléamainn was persued up *Gaic na Toiriche* (the gully of the persuer). Near the top he gave ground till

Muchcragan was close behind him, then turning rapidly, threw his battle-axe at the viking, cleaving his skull. Muchcragan was how-laid in a cairn which is still known as *Carn Mhuicragain*, and the mountain where it happened is *Beinn na h-Urchrach* (the mountain of the throw).

Cléamainn continued down through *Coire a' Mhulinn* (the corrie of the mill) where he gathered his family together and put to sea in his boat from near *Sgeir Cléamainn* (Clemant's skerry) and took refuge with the MacDonalds in Islay.

3. **Glac na Coimhridh (22). NM** ? G. *Glac a' Chòmhraidh* 'the hollow of the chatting'. *Glac* f. 'hollow, small glen'. *Còmhradh* 'conversation, chat'.

4. **Sgeir Chleamainn (22). NM** ? G. *Sgeir Chléamain* 'Clement's skerry'.

5. **Ba na fola (22). NM 5270 6216 (?).** G. *Bàgh na Fola* 'the bay of blood'. Described as a 'beautiful bay with towering cliffs west of Maclean's Nose'. *Bàgh* 'bay, cove'. *fuil* (gen. *fola*) f 'blood'.

6. **Allt an t-Salachair (26). NM 5150 6317.** G. *Allt an t-Salachair* 'the burn of the dirt'. Said to be the original name for the *Allt an t-Siùcair* 'the burn of the sugar', and allegedly renamed by *Alasdair Mac Mhaistair Alasdair* after having partaken of its sweet water.

7. **Glaic na Toiridh (26).** G. *Glaic an Tòiriche* 'the hollow of the pursuer'. *Glac* 'a hollow, small glen'. *Tòirich* 'pursuer'. The location is not known, but if the name refers to the shieling at **NM 5349 6412** then *an Toraidh* 'of the produce/milking' would be appropriate

Camas nan Clacha' Mòra

Corryvoulin and Skinad (detail)   National Records of Scotland, RHP72/1-8, Plan of Ardnamurchan and Sunart, Argyll, 1806.

Uamha na Creadha and an un-named cave in Camas Coire Mhuilinn; both were once inhabited

Coire Mhuillinn looking east to Sròn Mhòr and Maclean's Nose

GLEN BEG : Gleanna Beag : the small glen

1541, Glenbege (55); 1610, Glenbege (36); 1618, Glenbeg (8); 1667, Glenbeig (5); 1723, Glenbeg (3); 1784, Glenbeg (45); 2006, Glenbeg (7); 2007, Glenbeg (31).

18th/19th century settlement on Glen Beg

Valuation: Valued as a 2 merk land in 1541 and a 4 penny land in 1723.

Area: Of the 3994 acres recorded in 1806, there were 24 of ploughed arable, 21 cultivated with the spade, 3676 moor and pasture, and 273 acres of wood ground (7).

Tenants:
1541, Ferquhar McAlister McRanald (55); 1618, Donald McNeill VcConill Geir {or Yir}, Johnne McAllester VcIllespick, Johnne McCoull VcVicker, Johnne McConeill Voir VcFynvyne, Ewne McLauchlane VcConeill Geir (8); 1695, Eune Cameron and his children, Donald, Duncan and Mary; John due Cameron and his children, Donald and Duncan (38); 1716, Donald Cameron, Ewn Cameron his son, Alexander Cameron, their tenant, Alexander Cameron alias Iarlie, Donald McGresich, John Cameron, John McKenzie (37); 1737, Duncan Stewart (9); 1806, Capt. Alex. Cameron (7).

Population: 1723, 9 families, with 9 men, 11 women and 15 children, total 35.

Settlements: NM 586 628. Glenbeg. Twenty structures shown on the estate plan of 1807.
NM 5876 6234. The Mill.

Shielings: NM 583 655. At least twelve small structures on the west side of the burn and ten on the east side.

Placenames:

1. **Glen Beg (1). NM 586 628.** G.*Gleanna Beag,* literally 'the 'small glen'. This name appears to be transparent but there isn't a 'Small Glen' here. Glen Beg is that part of Glen More to the east of the 'Glenmore River'. Originally there could have been a single township here, *Gleann Mòr,* which was later divided into two, '*Gleann Mòr (Mòr)* 'the greater township of *Gleann Mòr*' and '*Gleann Mòr (Beag)* 'the lesser township of *Gleann Mòr*'.

2. **Lochan na Branaine (1). NM 6031 6382.** ?G. *Lochan nam Ban Uaine* 'the lochan of the pale women'? *bean* pl. *mnathan* 'wife, woman'. Named *Lochan na Mnatha Uaine* 'the lochan of the pale women' on OS maps. *Uaine* 'green, pale, wan, pallid, livid'.

3. **Port na Huagh (1). NM 5944 6143.** G. *Port na h-Uamha* 'the port of the cave' or with *uaigh* 'grave, tomb, den or cave'.

99

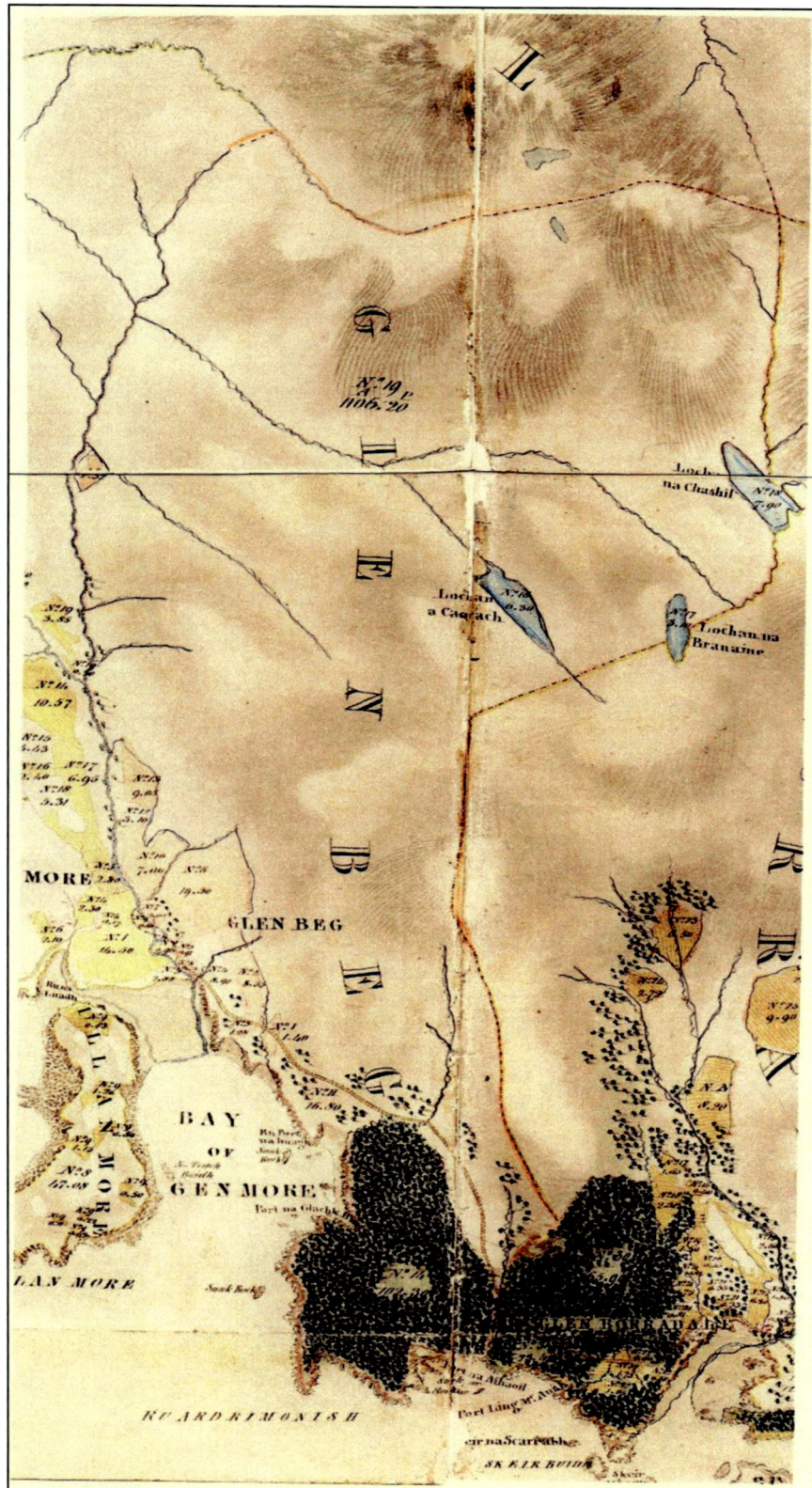

Glenbeg (detail)   National Records of Scotland, RHP72/1-8, Plan of Ardnamurchan and Sunart, Atrgyll, 1806.

4. **Port na Glachk Buige (1). NM 5960 6190.** G. ?*Port na Glaice Buige* 'the boat landing of the soft or boggy hollow'.

5. **Port na Athaoil (1). NM 5988 6084.** G. *Port na h-Àth Aoil* 'the port of the lime-kiln'.

GLEN DRIAN : Gleann (an) Droighinn : the glen of the thorn trees or brambles

1541, Glennyndriane (55); 1610, Glennyndriane (36); 1618, Glendreane (8); 1667, Glenindryane (5); 1722, Glendryen (3); 1723, Glendrain (3); 1737, Glendraen (9); 1739, Glendryan (13); 1784, Glendryon (45); 1806, Glendryen (7); 1828, Glendreyn (6); 2007, Glendrian (31).

The farmhouse, outbuildings and earlier dwellings

Valuation: Valued as a 2½ merk land in 1541 and 1610, and a 5 penny land in 1723.

Area: 1806, 2047.52 acres with 53.83 arable, 34.85 cultivated with the spade and 1958.84 moor & pasture.

Tenants: 1541, Lying waste (55); 1618, Allester McEan Voir VcEane (8); 1695, Neill McKendrie and his wife Finguell NcDonald with their children Eouin and John (38); 1716, John McColl, John McIllichreist, Donald McKenzie, Alexander Campbell, John McLachlan, Donald McLachlan (37); 1737, Duncan McEacharn (¼ of the 5 penny land), Patrick McIlvraw (¼), John McLachlan (⅛), Dougald McIllvraw (⅛), Donald McDonald (⅛) and Duncan Henderson (⅛); 1739, John McDonald (13); 1806, John McLauchlan and 5 others (7); 1828, 4 tenants on a year to year let (6); 1847, let to 7 tenants (46).

Population: 1723, 6 families, with 6 men, 8 women and 15 children, total 29.

Settlements: NM 4808 6872. Glendrian. Fifteen structures shown on the estate plan of 1807.

Shielings: NM 4682 7063. (Nine structures).   NM 6786 6831.
NM 4642 7018. (Six structures).   NM 4832 6902.
NM 4668 7057 (One structure).   NM 6829 6905.
NM 4484 7006 (Two structures).

Placenames:

1. **Glendrian (31). NM 4808 6872.** G. *Gleann (an) Droighinn* 'the glen of the thorn trees /brambles'.

2. **G. Coirabhan (?).** *An Coire Bàn* 'the white corrie'. Dative would be *(anns) a' Choire Bhàn*.

3. **Maol na Hullar (1). NM 4851 6990.** G. *Maol na h-Iolair* 'the hill of the eagle'. Named *Maol na Fhìr-eòin*, also 'the hill of the eagle' on OS maps.

101

Glen Drian National Records of Scotland, RHP72/1-8, Plan of Ardnamurchan and Sunart, Argyll, 1806.

4. **Lochan Donald Duich (1). NM 4995 6805.** G. *Lochan* 'Black Donald's Lochan'. Named *Lochan Mhic Dhonuill Dhuibh* 'the son of Black Donald's Lochan' on OS maps.

5. **Pynquhonych (55), Pitquhonicht (36); Pennyquhonich (52).** G. ?*Peighinn Chóinneach* 'the pennyland of the meeting place' or *Peighinn Choinnich* 'Kenneth's pennyland'. Valued as a 1 merk land in 1541, 1610 and in 1651, which makes it quite an important township, but the location has not been identified. It is listed between Glendrian and Faskadale, so it could be centered on the fields near Port Eigin-aig.

Early 19th century houses

The farmhouse abandoned c. 1942

Bealach Mòr and the Allt Mhic Cailean

Dykes round Field 17 (? Pynquhonych), ploughed in 1806, and right, within the main crofting area.

GLEN BORODALE : Gleann B(h)orradail : the valley of Borrodal

1541, Glenborvicdall (55); 1610, , Glenborviedall, Glenbervocdall (36); 1667, Glenbervidaill (5); 1686, Glenbadell (10); 1700, Glenboradill (10); 1700, Glenboridle (38); 1722, Glenboradale, (3), Glenboradil (20); 1723, Glenborradale (3), Glenbarrodale (4); 1742, Glenboradel (25); 1784, Glengoradell (45); 1806, Glenboradale (7); 2007, Glenborrodale (31).

Settlement site amongst the bracken

Valuation: Valued as a 2 merk land in 1610, a four penny land in 1723 and a five penny land in 1784 (45). In 1541, a one mark (2d) land held by Allester McAngus McKane and a xx s. (3d) land *fest till Ornisay* (Oronsay Priory) (55). Oronsay Priory was a monastic house dedicated to St. Oran and belonging to the Augustinian Priors Regular.

Area: Of the 1996 acres in 1806, there was 52 of ploughable arable, 26 cultivated with the spade, 1735 moor and pasture and 183 acres of plantations.
Glenboradale lies at the ffoot of a large Glen betwixt two high hills posessed by two brothers (half-pay officers) of the name of Cameron who have put up a pretty good house of stone & lime one story high thatch'd & two or three office houses of Drystone no arable ground about the house as it is all rocks but some pasture above the house about ¼ of a mile in the Glen in Different places are about 10 acres of arable ground the rest of the Glen very good pasture tho' rocky. (Anon, 1767).

Tenants; 1541, Allester McAngus McKane (55); 1686, Ewin Cameon (10); 1694, Alexander and Finguell Cameron, and their children Eouin, Donald elder and younger, and Allen (38); 1697, Eune and Elizabeth Cameron (38); . 1697, Eune and Mary Cameron with their children, Finuill and Katherine (38); 1716, Allan Stewart, Donald Stewart, John glass McColl, Ewn McIllemichell, Duncan McColl na Cronie, Donald McLachlan, John McColl, Ewn dow McColl, Duncan McColl, John Stewart and John McInnish riach (37); 1740, Sanders Cameron (53); 1742, Allester mc Allan vic Eun vic Allester (25); 1806, Capt. Alexr. Cameron (7); 1828-1837, Allan Cameron (6).

Population: 1723, 6 families with 9 men, 13 women and 11 children, total 33;

Settlements: NM 6120 6137. The fragmentary remains of two (or more) turf houses and other structures.

Shielings: NM 6153 6238.
NM 6153 6233.

Glenborodale

National Records of Scotland   RHP72/1-8, Plan of Ardnamurchan and Sunart, Argyll, 1806.

Glenborodale Bay

Placenames:

1. **Glenborrodale (31). NM 6105 6110.** G. *Gleann Bhorradail* 'the glen of *Borrodal'*. ON *Borg-dalr* 'Fort Valley'. *borg* 'fort', and *dalr* 'valley'.

2. **Port Ling McAush (1). NM 6036 6074.** G. *Port Luinge 'McCuish* 'the port of McCuish's ship'?

3. **Skeir na Scarrabh (1). NM 6045 6055.** G. *Sgeir na Sgarbh* 'the skerry of the cormorant'. Named *Sgeir Bhuidhe* 'the yellow skerry' on OS maps.

4. **Loch na Brachk Beg (1). NM 6100 6440.** G. Perhaps *Loch nam Breac Beag* 'the loch of the little trout', with *breac* 'trout'.

5. **Loch na Brachk More (1). NM 6100 6465.** G. *Loch nam Breac Mòr* 'the loch of the big trout'.

6. **Maol Corrie Loch Choshil (1). NM 6068 6558.** G. *Maol Coire Loch a' Chaisil* 'the bare hill of the corry with the loch of the mound'. *Caiseil* 'wall, mound'. The OS has it fem. *na Caisil,* but this may be local usage.

7. **Ault Maol Na Neach (1). NM 6273 6316.** G. *Allt Maol nan Each* 'the bare hill of the horses burn'.

Risga, 'Diavan Risgay', Sgeir Buidhe and Sgeir a' Choir

8. **Diavan Risgay (1). NM 6074 6006.** G. ? *Dìomhain Risga* 'useless Risga'? *Dìamhain* 'trifling, useless'. A rocky islet off the west side of Risga.

106

9. **Innis nam Feorag (19).** NM ?? G. *Innis nam Feòrag* 'the meadow of the squirrels'. Alasdair Mac Shomhairle used to see a *Glaisirig,* a fairy, washing herself on a rock up below the bridge here.

10. **Illan na Neah (24). NM 6100 6050.** G. *Eilean an Fhèidh* 'the island of the deer'. *fiadh (*gen. & pl. *fèidh* 'deer'.

11. **Change House (?). Tigh an Rudha (26). NM 6224 6054?** G. Possibly *Taig-òsda,* but the inn may always have been referred to by the English 'Change House'. *Taigh an Rubha,* 'the house of the point'. This appears to have been an inn close to Glen Borrodale Castle and the Carna Ferry. In 1870, Mary MacPherson of Smirissary was accused of malicious mischief by setting fire to heather and destroying fir plants, the property of James Dalgleish of Ardnamurchan. The accused had set fire to the heather while making a *smuidean,* or smoke-raising, to attract the ferryman to cross Loch Sunart to take her to Glencripesdale (48).

12. **Creagnish (18). NM 6188 6713.** Named *Creag an Fhithich* on OS maps, and Creagnish may be a phonetic rendering of this. It translates as 'Raven Crag'.

13. **Drochaid na h-Innse (19).** G. *Drochaid na h-Innse* 'the bridge of the meadow. *Drochaid* 'bridge'. *Innis* (gen. *innse*) f 'meadowland'.

14. **Eaglais Choinnich or Kenneth's church' (26). NM 618 607.** During the mid 19th century the Presbyterian Minister of the Gospel was Kenneth MacKenzie, a native of the Kilchoan district, who used to hold services here, although during his ministry he resided at Camasinas.

The dùn called Caisteal Breac

Glen Borrodale from Caisteal Breac

107

GLENMORE : An Gleann Mòr : the great glen

1541, Glennmoir (55); 1610, Glenmoir (36); 1667, Glenmoir (5); 1722, Glenmor (3), 1723, Glenmoir (4); 1784, Glenmore (45); 1806, Glen more (7); 2007, Glenmore (31).

Eilean Mòr from Glen Beg

Valuation: Valued as a 2½ merk land in 1541, and a five penny land in 1723.

Area:
From Glenborodale to Glenbeg & Glenmore who both ly at the foot of the Mountains only separated by a burn are inaccessible rocks with a few fine oak, ash & some birch interspers'd. At the back of the hills are large glens full of Rocks & moss Pasture but no wood – The Glens above the Houses are Extensive & fine pasture and about the houses which good & built of dry stone & thatch'd are about 10 acres of arable ground pretty good soil of a Clay kind, and below the Houses betwixt them and the sea ly about 5 acres of Low or Meadow Ground of a good soil & fine grass thro which the burn runs. (Anon, 1767).

Tenants: Early 1500s, Farquhar McAlister McRanald (Exchequer Rolls; (26): 1691, Gillies McGillies and his wife Katherine Campbell (38); 1693, Donald McGillies and his wife Katherine McIleis (38); 1708, Eun Cameron (10); 1716, Donald oig McKendrig, Duncan McKendrig his son, Gilmichael McCallum (37); 1737, Duncan Stewart (9); 1806, Alexander Cameron and 12 others (7); 1825-1846, 8 tenants (6); 1828, 8 tenants (6); 1847, 8 tenants (46).

Population: 1723, 5 families, with 9 men, 7 women and 13 children, total 29.

Settlements: NM 5854 6230. Glenmore. Eleven structures shown on the estate plan of 1807.
NM 585 628. Nine structures in 1807 (1).

Shielings: Not recorded.

Placenames:

1. **Glen More (1). NM 585 623.** G. *An Gleann Mòr*, 'the greater township of *An Gleann Mòr*'. See 'Glen Beg' .

2. **Bay of Glenmore (1). NM 5900 6159.** 'The bay of the great glen'. Now **Glenmore Bay.**

3. **Ru na Luadh (1). NM 5860 6194.** G. Probably *Rubha na Luath* 'the headland of the ashes' or with *luaidh* 'lead or lead shot'.

4. **N'Teach Buidh (1). NM 5892 6114.** G. *An t-Each Buidhe* 'the yellow horse'. Named *Sgeir an Eich Bhuidhe* 'the skerry of the yellow horse' on OS maps. The rock is covered with yellow seaweed.

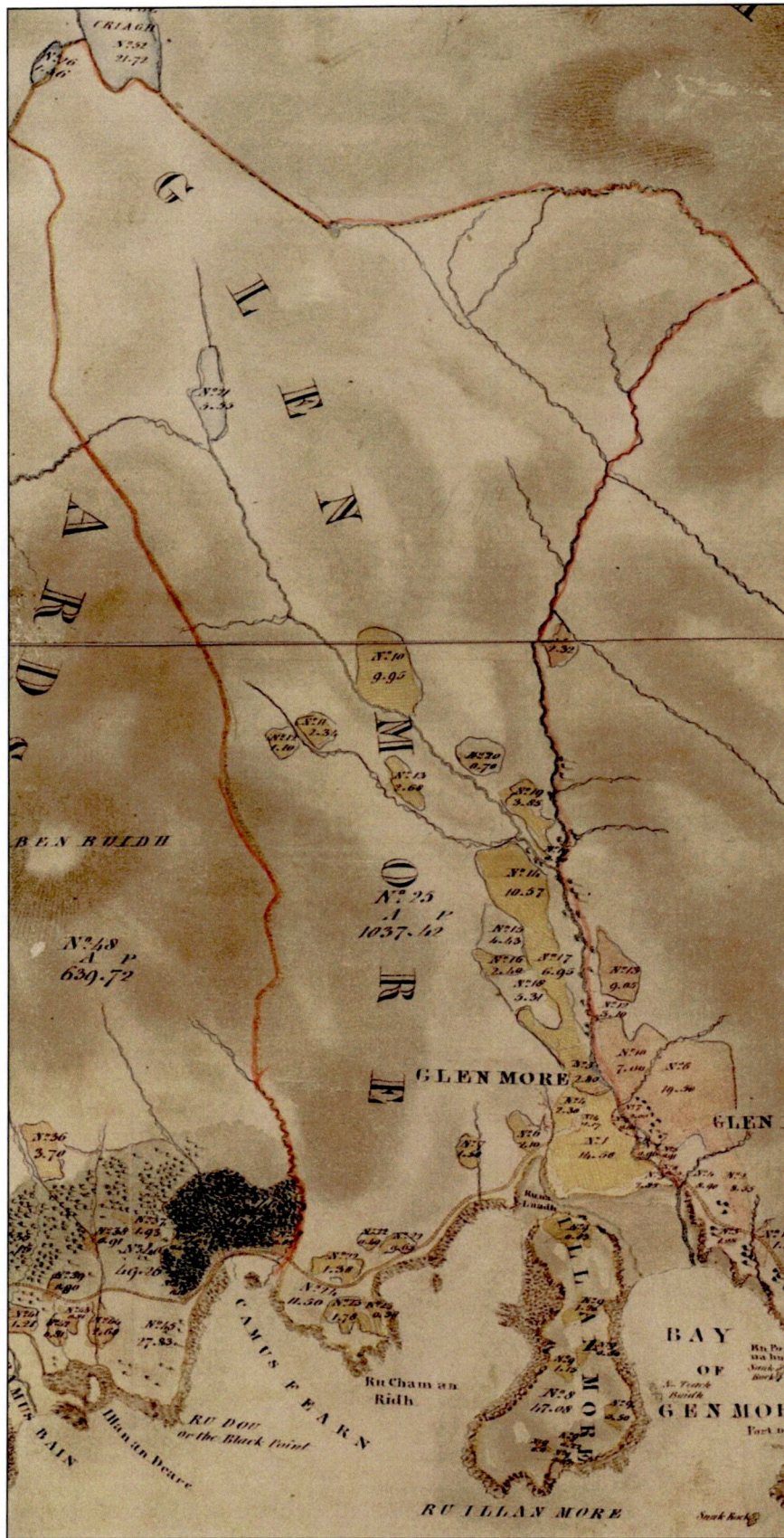

Glen More National Records of Scotland, RHP72/1-8, Plan of Ardnamurchan and Sunart, Argyll, 1806.

GRIGADALE : Griogadal : the gravelly valley

1541, Geredale (55); 1618, Girgodill (8); 1686, Girgidill (10); 1723, Girgadil (3); 1723, Girgadell (3); 1737, Girgaddale (9); 1767, Girgadale (29); 1784, Girgadell (45); 1806, Grigadale (7); 2007, Grigadale (31).

Loch Grigadale

Valuation: A 2½ mark land in 1541 and a 5 penny land in 1723.

Area: Of the 1561.92 acres in 1806, 16.26 were arable, 64.61 cultivated with the spade and 1481.05 moor & pasture.

Girgadale lies in glen with about 15 or 20 acres of arable or grass ground about it. The Houses such as were mentioned above (pretty {good?}being built with Stone & ffail) all of a light red soil They use here by way of Manure a White Shell kind of sand which is in great plenty on the Coast. To the South of the Town lies a small Loch of the same name ffamous they say for Trout, & in the Burn that runs from it some pearl muceles, a great deal of hill pasture round it but the arable ground cannot be greatly enlarged. The Coast here is said to abound with good grey ffish of all kinds & to the west of this is a fine station for a Harbour & fish town (Anon, 1767).

Part of the Moor Pasture of this Farm is wet, which subjects the Cattle on it to a Disease called the Cruban, and the Sheep to the Rot. It besides is exposed to Sea blasts from all points but the East. Alex. Lowe, 1807.

'Cruban (or Crutching) is occasioned by eating strong rough grafs by which they are so much bound up in their bones and joints that they can scarcely travell about and so lean that their bones are like to come thro' the skin – if they get good grafs when the frost comes on they recover' (46).

Tenants: 1541, set to Ycomkill (Iona Abbey) (55); 1618, Alexander McInnes VcEane and Dougall McGillivorie (8); 1686, Ronald VcDonald VcEan (10); 1693, Donald McGressich, his wife Christian and sister Mary (38); 1693, Finla McGreeick, his wife Katherine NcIlvrae and their daughter Katherine (38); 1694, Ronald McDonald and his son John (38); 1699, John McNiven and his wife Mary McDonald. Her brothers and sister were John, Archibald, Alexander, Ronald and Katherine (38); 1716, John McIlvra, John McNiven, Donald McKendrig, workman, John dow McKendrig, Angus McIlveoll, Neil McBrian (37); 1737, fflorence McDonald, relict of John McNiven (9); 1739, James Stewart (23); 1807, James Campbell and one other (7); 1828, three brothers, Alex, Donald and John Cameron on a year to year tenancy (6).

Population: 1723, 5 families, with 9 men, 8 women and 7 children, total 24.

Settlements: NM 4325 6740. *Grigadale.*

Grigadale (detail)   National Records of Scotland,,RHP72./1-8, Plan of Ardnamurchan and Sunart, Argyll, 1806.

Shielings: NM 4165 6609. 'Port Min'. House and field.
                   NM 4251 6527. 'Glac nam Muileach'. Developed shieling with field system.

Placenames:

1.   **Grigadale (31). NM 4325 6740.** G. *Griogadal* 'the gravelly valley'. A loan-name from ON –*dalr* 'valley'.

2.   **Port Nian (1) NM 4165 6625.** ?G. *Port nan Eun* 'the boat landing of the birds'. Named *Port Min* 'the smooth landing place' on OS maps, with *Port Garbh* 'the rough landing place' some 300m to the SSW.

3.   **Ru Corrach a' More (1). NM 4140 6625.** ?G. *Rubha Corrachaiche Mòr* 'the big headland of looking out'. Named *Corrachadh Mór* 'the abrupt headland' on OS maps, which is probably a better interpretation.

4.   **Lochan na Clash (1). NM 4265 5435.** G. *Lochan na Clais* 'the lochan of the long hollow'. Named *Lochan Druim na Claise* 'the lochan of the ridge of the long hollow'. on OS maps. *Clais* gen. *claise* f 'a ditch, trench, gully or long hollow'.

5.   **Achdraisaid (1). NM 4332 6358.** G. *An Acarsaid,* and named *An Acairsaid* 'the harbour' or 'the anchorage' on OS maps. *Acarsaid* f. 'an anchorage, harbour, mooring'. The early 19th century spelling suggests a field name, but the nearest fields are at *Reidh-dhail,* 1.9km to the NNE.

111

6. **Illan a Vanach (1).** NM 4245 6800. G. *?Eilean a' Bhànaich* 'the island of the whitening'. Was this a bird-rock made white with droppings during the summer months ? Named *Eilean Carrach* on OS maps. *Carach* adj. 'twisting, changeable, unreliable'. *Carrach* adj. 'stony, rocky'. Possibly *Eilean Mhanaich* 'Monk's Island'. Perhaps a hermit lived there at some time. The adjacent and equally inhospitable island is *Eilean Chaluim Cille* 'St. Columba's Island'. If correct, then both names may reflect the fact that Grigadale was held by Iona Abbey in 1541.

7. **Corra Bein (1).** NM 4300 6781. G. *A' Chorra Bheinn* 'the steep hill'. Named *Sgurr nam Meann* 'the rocky peak of the kids' on OS maps.

8. **Port na Carra (1).** NM 4236 6736. ?G. *Port na Corra* 'the port of the heron ?'.

9. **Avhin Grigadale (1).** NM 4280 6710. G. *Abhainn Griogadail* 'River Grigadale'. Named *Allt Grigadale* 'Grigadale burn' on OS maps.

10. **Eilean Nach Freobhadh Aon Scisireach (22).** NM 4246 6392. G. *An t-Eilean nach Treobhadh Aon Seisreach* 'the island one team couldn't plough'. Named **Illan na Seachd Seisrichin** or '**The Island of the Seven Ploughs**' on RHP72 and *Eilean nan Seachd Seisrichean* 'the island of the seven (plough) teams' on OS maps.

    Many years ago, one of the tacksmen of Grigadale decided that it was time he got married, but none of the local ladies were attracted to him as he was violent and domineering. He learned of a lady in Mull who might make a suitable wife, and when he approached her, she expressed a willingness to marry him providing her father agreed. Her father was not impressed until the Grigadale man told him that he had an island that seven teams of horses could not plough even if they worked at it for a full year. Assuming that the Grigadale man was prosperous and could provide for his daughter, he readily agreed to their marriage.

    When the happy couple returned to Grigadale the good lady asked to see the wonderous island. Walking through the township to *Lochan Druim na Claise*, at a point overlooking *Port Choinnich,* they could see the island some 400 feet below, a small, barren rock, where no plough team had ever landed, and probably never will. Who the tacksman was is not recorded, but he was probably a MacNiven, a family of incorrigible rogues who dominated West Ardnamurchan in the first half of the 18th century (57).

11. **Eilean Redall (23).** NM 4246 6392. G. *Eilean Reidh-dhail* ' Redall Island' and not 'the island of the smooth field'. *Rèidh* adj. 'clear, smooth, even'. *Dail* f. 'dale, meadow, valley', but here, a 'field'. The same island as No. 10, which is a low, barren rock.

Bàgh MacNeil

12. **Bàgh MacNéill (17).** NM 4260 6780. G. *Bàgh MhicNéill* 'MacNeill's Bay'. This is where the MacNéills of Barra are said to have landed their cattle for onward transmission to the great fairs of the Crieff and Falkirk Trysts. It is hard to imagine a more impractical route.

KILCHOAN : Cille Chomhain : the church of St. Còmgan

1541, Kilzeorquhanne (55); 1610, Kilyeorquhane (36); 1618, Kilchoane (8); 1639, Killhoan (37); 1667, Killechnane (5); 1695, Kilchoan (38); 1722, Kilchoan (3); 1723, Kilhoan (3); 1767, Killyhoan (29); 1784, Killocgoan (45); 1847, Kilchoan (46); 2007, Kilchoan (31).

The old church and burial ground

Valuation: Valued as a 2 merk land in 1610 and a five penny land in 1723.

Area: *S.W from Skinned at the distance of about 2 Miles on the side of a hill fronting the sea lies the Kirk of Killyhoan which was lately rebuilt & Slated & about it a Burial place but not inclosd (in 1767). Betwixt it & the sea lies a small Farm of the same name with a very good Millhouse of stone and lime & slated about which are 5 or 6 acres of grass ground & 6 or 7 of Meadow or low ground betwixt it & the sea, included in which is the Minister's Glebe which he has let at £5 per an. There is also a pretty good house on the Farm built for a publick house by Sir Alexr Murray – There is also some Limestone on this Farm-all these Farms is to be understood to have very large & extensive hill pastures there is a great deal of Ware all along this coast for Kelp. The Tennants use a good deal of it for that purpose as also for Manure for their ground, but there is no Fishery of any kind but some that the Tennants catch for their own use and by all accounts there seems to be very little Encouragement for erecting any thing of that kind (Anon,1767).*

Tenants: 1541, Donald McAlister McKane (55); 1618, Ewne McFinla VcGillene [or VcGillevir] and John McFinla his son (3); 1694, John McCruter and his wife Ann NcDonald (38); 1716, Archibald McIlehavich, Dugald, Archibald and John McIllehavish his sons, John Campbell, Donald roy McBrion, Donald McBrion, Lachlan McBrion, Donald McCallum, John McCallum, John dow McBrion (37); 1737, Alexander McDonald (9); 1739, Donald McKenzie, John McColl, Duncan McKenzie and Donald McIllyvrie (13); 1806, Donald McKenzie and 5 others (7); 1828, 8 tenants on a year to year let, including D. McKenzie who also had a 19 year lease of the Public House (6); 1847, 12 tenants (46). 1852, 6 tenants were evicted for non-payment of rent (53).

Population: 1723, 8 families, with 13 men, 12 women and 10 children, total 35.

Settlements: NM 4810 6384. Kilchoan. Mill and 5 houses on the estate plan of 1807.
NM 486 639. Three houses in 1807.

Shielings: Not known.

Kilchoan

National Records of Scotland, RHP72/1-8, Plan of Ardnamurchan and Sunart, Argyll, 1806.

114

Placenames:

1. **Kilchoan (31). NM 489 638.** G. *Cille Chomhghain >Cille Chomhain* 'the church of St. Còmgan'. Còmgan founded a monastery in Lochalsh c. AD 673, and his feast day is 13th July. This was an Early Christian site and one of the two mediaeval parishes of the district, the other being Island Finnan in Loch Shiel.

2. **Maol na Bohan (1). NM 4887 6648.** G. *Maol nam Bothan* 'the great bare rounded hill of the shieling huts'.

3. **Maol Chorrie Chruinn (1). NM 4924 6632.** G. *Meall a' Choire Chruinn* 'the bare hill of the round corrie'. Named *Meall an Tarmachain* 'the bare hill of the ptarmigan' on OS maps.

4. **Maol na Shomraig (1). NM 4920 6585.** G. *Maol an t-Samhraidh* 'the bare hill of summer', 'the summer grazing hill', or with *samhrag/seamrag,* probably 'bird's foot trefoil' in this location.

5. **Skerr na Carraidh (1). NM 4810 6314.** G. *Sgeir na Cairidh* 'the skerry of the fish trap'. *Cairidh* (gen. *cairidhe*) f. 'fish trap'.

Glas Eilean from the old kirk

6. **Sailean Chill-Chomhain (2). NM 4845 6350.** G. *Sàilean Chille Chomhain* 'the inner reaches of Kilchoan Bay'. Translated on the OS maps as 'Kilchoan Bay'.

7. **Glebe (1). NM 4840 6410.** The Estate Plan of 1807 shows the Glebe Croft to consist of 22.64 acres of moor and 10.16 acres of arable land under the plough.

8. **Mill Croft (1). NM 4814 6388.** The Estate Plan of 1807 shows the croft to consist of two ploughable fields measuring 3.25 and 0.55 acres.

9. **Inn Croft (1). NM 4808 6384.** The Estate Plan shows this croft consisting of two ploughable fields measuring 1.15 and 2.65 acres.

10. *10.* **Bou na Keill (1). NM 4810 6314.** G. *Bodha na Cille* 'the sunk rock of the church'. A dangerous sunk rock in the middle of the bay, and only seen at low water. Possibly the same as *Skerr na Carraidh (* No. 5).

11. **Bodha Ruadh (34). NM 482 6355.** G. *Am Bodha Ruadh* 'the red sunk rock'.

12. **Tom na Cille (34). NM 4831 6424.** G. *Tom na Cille* 'the knoll of the church'.

13. **Allt a' Mhuillinn (34). NM 4817 6367.** G. *Allt a' Mhuillinn* 'the mill burn', which is the (English) name on the OS maps. The remains of the mill, with the lade and some machinery can still be seen on the Kilchoan side of the burn. David Grant submitted a detailed estimate for refurbishing the mill towards the end of 1833. The mill wheel was to be 11 feet in diameter and 2½ feet wide, with a 10 feet square corn drying kiln. It was completed by 1 February 1834, four months after the commencement date and cost £110-0-0. (GD241 Box 64/2112).

14. **Cnoc na Gour (1). NM 4490 6330.** G. *Cnoc na Gobhar* 'the hill of the goat'.

The cairn on *Tom na Cille* looking down the Sound of Mull

Windows in the Old Parish Church of Kilchoan

116

KILMORY : Cille Mhoire: St. Mary's church

1541, Kilzemory (55); 1610, Kilyemory (36); 1667, Kilzemore (5); 1722, Kilmory (3); 1723, Kilmorrie (3); 1767, Kilmory (29); 1784, Kilmorie (45); 1828, Kilmory (6); 2007, Kilmory (31).

Croft house on Kilmory

Valuation: Valued as a 2½ merk land in 1541 and 1610, a four penny land in 1723 and a five penny land in 1784 (45).

Area: In 1806 there was 53.08 acres of arable land, 82.29 cultivated with the spade and 776.27 of pasture, a total of 911.64 acres.

Tenants: 1541, Johne Erenoch (*an Eireannaich*, the Irishman?) (55); 1716, John McKendrick VcEwn dui, John dow McKenrig, Pat McIlharish, Angus Mc Brionn, John McEan oig McEachern, John McNiven, Archibald McDonald, John McDonald, Dugald Johnstoun, and his son John, Duncan Mcpherson, John McKuarige (37); 1737, Archibald McNiven, who was also the tenant of Swardle Chorrach (9); 1828, ten tenents (6).

Population: 1723, 6 families, with 10 men, 10 women and 8 children, total 28.

Settlements: NM 5266 7037. Kilmory.

Shielings: NM 5478 6721. 'Allt nan Coireachan'.

Placenames:

1. **Kilmory (31). NM 529 701.** G. *Cille Mhoire* (or *Cille Moire*) 'St. Mary's church'.

2. **Maol na Corinan (1). NM 5524 6719.** Named *Meall nan Coireachan* 'the rounded hill of the corries' on OS maps, which may be the correct Gaelic.

3. **An Tobar Baistidh (2). NM 5313 6999.** G. *An Tobar Baistidh* 'the christening font'. A small font in the old burial ground, said to be never empty of water.

4. **Druim na Croise (2). NM 5300 7033.** G. *Druim na Croise* 'the ridge of the cross'. *Crois* gen. *croise* f. 'a cross'.

5. **Tobar Mairi (2).** G. *Tobar Moire* 'St. Mary's Well'.

Kilmory Burial Ground

6. **Meeting House (1). NM 5313 6999.** This may have been the old church inside the burial ground. When Swordilcheil and Ockile were let in 1732, the tacksmen were obliged to 'assist the other tenants by turn in thatching the **Sermon** or **Meeting House** of Kilmory'. Latterly it was at NM 5319 6989, and recently converted into a dwelling house.

7. **Port Kilmory (15). NM 5265 7053.** G. *Port Cille/Chille Moire* 'Port Kilmory'. 'the port of St. Mary's church'. This was where bodies destined for burial in Kilmory were brought in by sea.

8. **Illan Aird Toe (15). NM 5268 7122.** G. *Eilean Àird Tobha,* with *tobha* 'headland', 'the island of the high headland'. Named Ardtoe Island on OS maps, with the same meaning. A height attached to the mainland by a long, low flat piece of land, the height in question appearing to be at the end of a tow rope. The 'island' is a considerable expanse of slabs covered by the sea at flood tide, with about ten skerries above high water, but hardly an 'island' or a 'height'.

9. **Ru na Faolin (15). NM 5265 7122.** G. *Rubha nam Faoileann* 'the headland of the common white gulls'. The OS records this headland to the east, at NM 5314 7094. Hugh MacKenzie refers to it as *Rubha Dubh na Faoileann* 'the black headland of the gulls'.

10. **Imire (21). NM 5305 7004.** G. *An t-Imire* 'the field' or 'the ridge of land'. 'In the middle of the field, and to the north-west of it (the burial ground), is a small meadow known as *'Imire'* (land of the dead). When a funeral came by boat, perhaps from as far north as Arisaig, the body was conveyed from the point where it was landed at the shore straight to this field, and here the procession would halt before proceeding to the graveyard, a few yards distant'.

11. **Crois ni Mhath (21). NM 5365 6995.** G. *Crois Nì Mhath* 'the cross of God'. *An Nì Math* 'God'. *'Crois nì Mhath* is the name given to the ridge of hill to the north-east of the churchyard. Travelling across country from that direction mourners would have their first glimpse of the churchyard from the top of this ridge. They would stop there and cross themselves; hence the name, which means 'The cross that would do good'. Hugh MacKenzie suggested that this is where people of the Roman Catholic faith stopped to pray on their way to church. Was there once an actual cross here?

118

12. **Ru Ghari Lea (1, 47). NM 5234 7105.** G. *Rubha a' Gharaidh Lèith* 'the promontory of the grey cave'. *rubha* 'a promontory', *garaidh* 'cave or den', *lèith,* gen. of *liath* 'grey'. Alternatively G. *Rubha Garadh Liath* 'the headland of the dyke, rocky enclosure, lambing pen or garden', but none of these seem plausible here.

13. **Cistichan (1). NM 5300 7088.** G. *Na Cisteachan* 'the chests, boxes or coffins'. The name appears along the top of the cliffs at this point; does it suggest a number of prehistoric burial cists ? Hugh MacKenzie (47) places *Cisteachan* on Swardle Chorrach in a more likely position.

14. **Loch an Gilet (1). NM 5562 6700.** Named *Lochain nan Gillean* 'the lochan of the young men' on OS maps.

15. **Drum Scrile (1). NM 6425 6889.** G. *Druim Sgrial* 'the ridge of the scree'. Named *Druim an Scriodain* - with the same meaning- on OS maps. The same combination of *Sgrial/Scriodain* occurs in Sallachard (3), Ardgour.

16. **Braigh (47). NM 5295 7007.** G. *Braighe* 'the upper part (of the township)'.

17. **Cean Ishel a' Bhaile (47). NM 5286 7035.** G. *Ceann Iosail a' Bhaile* 'the lower end of the township'.

18. **Balnaha (47). NM 5297 7060.** G. *Baile na h-Àtha* 'the township of the kiln'. Named 'Balnaha' on OS maps. The generic term *Baile* 'town' is unusual throughout the district, and generally refers to the principal settlement on a township. Hugh MacKenzie emphasised the pronunciation *Baile na h-Atha*, so it is <u>not</u> *Baile an Àth* 'the township of the ford'.

19. **Tor Shollisht (47). NM 5300 7035.** G. *Tòrr Sholuis* 'the hill of light'.

20. **Uamharach na Marbh (47). NM 5264 7045.** G. *Uamharrach na Marbh* 'the gloomy (literally terrible) (path) of the dead people'. This is the route the Arisaig and Moidart people used to take the deceased from the shore to the Kilmory burial ground.

21. **Allt a' Harstcha** or **Allt a' Farsha/Charste (47). NM 5320 7035.** Possibly G. *Allt a' Chaise* 'the burn of the cheese'.

22. **Cnoc na Callum (47). NM 5313 7010.** G. *Cnoc na Calman?* 'the hill of the pidgeon'?

23. **Gorstan an Daimh (47). NM 5297 7068.** G. *Gortan nan Damh* 'the field of the stags'.

Rock Stack in Port Kilmory

24. **Port Beag (47). NM 5238 7077.** G. *Port Beag* 'the little boat landing'.

25. **Bogha Laogh (47). NM 5243 7121.** G. *Bogha na Laogh* 'the sunk rock of the calf'; 'the calf skerry'. Possibly a small skerry in close proximity to a point which might resemble a cow.

26. **Lacht Biorach (47). NM 5235 7110.** G.? *Leac Biorach* 'a sharp ledge of rock jutting out into the sea and covered by the incoming tide'.

27. **Uamha Calaman (47). NM 5315 7085.** G. *Uamh a' Chalamain* 'the cave of the dove'.

28. **Ben Toochda (47). NM 5274 7035.** G. *?Beinn na h-Uachdair* 'the grazing land with the farm stock', with *beinn* 'pasture or grazing land' as in Barra and South Uist, and *uachdair* 'farm stock'.

29. **Tobar nan eun (47). NM 5284 7023.** G. *Tobar nan Eun* 'the well of the birds'.

Croft house at Kilmory with a painting on the stable door, and a rock stack in Port Kilmory

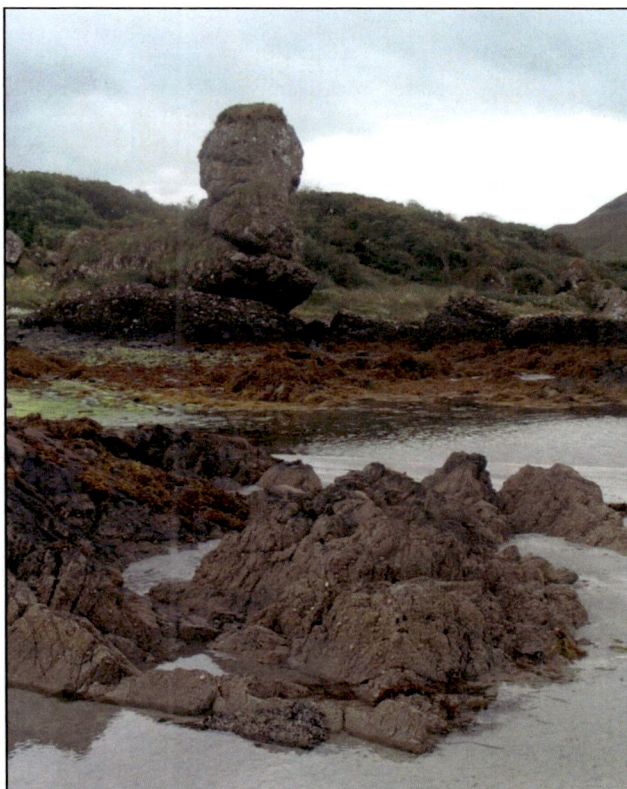

LAGA : An Lag Àth : the ford hollow

1541, Lawga (55); 1610, Lawga (36); 1618, Lagre (8); 1667, Langa (5); 1686, Lagan (10); 1722, Laga (20); 1767, Laga (29); 1827, Laga (6); 2007, Laga (31).

Camas Bàn, Laga Bay

Valuation: Valued as a 2½ merk land in1541 and a 5 penny land in 1723.

Area: In 1807 the 1653.16 acres were made up of 55.10 acres of ploughable land, 29.08 cultivated with the spade, 99.33 pasture, 28.60 lochs, 48.15 planted woodland and 1392.90 moorland.

Tenants: 1541, Ewin McGownele McCauchir and Gillaspy Roy (55); 1686, Donald Cameron (10); 1716, Duncan Stewart, John McIllmichel and John dow McColl his servants, Archibald McEan vuy, Archibald Stewart, maltster, Gilbert McIntyre, Duncan dow McLachlan, John McEan vuy, a herd boy and Duncan roy McKenzie (37); 1727, 9 tenants, 8 shares (6); 1737, Alexander Cameron, along with Camasinnes (9); 1760, Alan Cameron (10); 1807, John Mc.dougal and 8 others (7); 1828, 9 tenants (6); 1847, 10 tenants (46).

Population: 1723, 7 families, with 13 men, 8 women and 11 children, total 32.

Settlements: NM 6335 6105. Laga. Fifteen structures shown on the estate plan of 1807.

Shielings: NM 6367 6198 Five drystone footings.
NM 6324 6163. One probable shieling hut.
NM 636 619 Five structures in the vicinity.

Placenames:

1. **Laga (31). NM 6335 6105.** G. *An Lag Àth* 'the ford hollow'. Possibly from ON *Lága* 'a hollow' or *Lágr* 'low'.

2. **Maol Lochan a' Loist (1). NM 6321 6432.** G. *Maol an Lochain nan Losgann* 'the hill of the lochan of the frogs (or toads). Named *Meall nan Each* 'the hill of the horses' on OS maps. *Maol an Lochain Loisgte* 'the hill of the burnt lochan' seems unlikely.

3. **Ault na Luib Rioch (1). NM 6437 6400.** G. *Allt na Lùib Riabhaich* 'the burn of the speckled bend'. *Lùb* gen. *lùib* f 'a bend, loop'. *Riabhach* 'spotted, stripy' or 'grey, brown, drab'. This is the burn running north from Loch Laga, forming the upper reaches of the *Allt Beithe*. **An Lub Riabhach (54). NM 6440 6414.** G. *An Lub Riabhach* 'the speckled bend'.

4. **Ault Rie Loch Laga (1). NM 6386 6672.** G. *Allt Ruidhe Loch Laga* 'Loch Laga shieling burn'. *Ruighe* 'a hillslope, a shieling'.

5. **Ault Rie na T'lugan (1). NM 6381 6322.** G. *Allt Ruidhe an t-Slugain,* 'the shieling of the gorge burn'. *Ruighe* 'hillslope, shieling'. *slugan* gen. *slugain* m. 'gorge'.

121

Laga

National Records of Scotland, RHP72/1-8, Plan of Ardnamurchan and Sunart, Argyll, 1806.

6. **Ault Derry na Haclaish (1)**. NM 6480 6100. G. *Allt Doire na h-Achlaise* 'the burn of the armpit grove'. *Doire* f. 'a clump of trees, thicket'. *achlais* gen. *achlaise* f . 'armpit, bosom'.

7. **Ault Corrachan Duinn (1)**. NM 6449 6150. G. *Allt a' Choireachain Duinn* 'the burn of the brown corries'. *Coireachan* pl. of *coire* 'corries'. *Donn,* gen. *duinn* ' brown'.

8. **Port Stron na Brathnan (1)**. NM 6265 6065. G. *Port Sròn na Bràthntan* 'the port of the point of the quern'. 'the port of *Sròn na Bràthan'* 'of the quernstone' or *'Sròn nam Brà'* 'of the quernstones'.*bra* pl. *bràthntan* 'quern'. Recorded as being a source of quernstones in 1723 (4).

9. **Ru Lean (1)**. NM 6209 6037. G. *An Rubha Leathan* 'the broad headland'.

10. **Port More (1)**. NM 6448 6024. G. *Am Port Mòr* 'the big port'.

11. **Coire na Gael (18)**. NM 6322 6180. G. *Coire nan Gàidheal* 'the corrie of the Highlanders'.

12. **Breun Torr (18)**. NM 6400 6078. G. *Am Breun Thòrr* 'the dirty hill'. *Breun* (gen. *brèine*) f. 'stench, stink, filthy'.*Breun* usually aspirates the noun, hence *thòrr* and not *tòrr,*

13. **Stron na Buirate (18)**. NM 6284 6063. G. *Sròn nam Buairte ?* 'the promontory of the troubled ones'. *Buairte* 'troubled, tempted', although this seems unlikely. Named *Sron nam Bràthan,* 'the promontory of the quernstones' on OS map. There are broken and unfinished querns on the shore and in the rocks here, and Murray of Stanhope noted that in 1723, Laga produced 'excellent miln stones'. Possibly *Sròn na Bràthan* was intended here.

14. **Port na Chellestar (1)**. NM 6418 6037. G. *Port an t-Seileastair,* 'the port of the iris' (or sedge).

15. **Maol na neach (1)**. NM 6312 6373. G. *Maol nan Each* 'the hill of the horses'. Named *Leac an Fhidhleir* 'the slabs of the fidddler' on OS maps. *Fidhleir* s.m. 'fiddler, violinist'.

16. **Allt Maol na neach (1)**. NM 6268 6368. G. *Allt Maol nan Each* 'the burn of the hill of the horses'.

17. **Lochan na Achruimleid (1)**. NM 6284 6235. G. *Lochan a' Chrom-leathaid.* 'the lochan of the crooked hill slope' as on OS maps.

18. **Lochan Beinn Laga (54)**. NM 6437 6232. G. *Lochan Beinn Laga* 'the Ben Laga lochan'. This was on a piece of land that was in dispute between the Laga and Camasinas tenants in 1806

Loch Laga

19. **Allt Lochan Beinn Laga (54)**. NM 6400 6238. G. *Allt Lochan Beinn Laga* 'the Lochan Ben Laga burn'. The burn running west out of the lochan into the *Allt Mór.*

Laga fields from the south-east

Laga fank

Laga fields looking east

One of the Laga shielings

MINGARY : Mìogharraidh or Mikinn Garðr: the great enclosure

1541, Meare (55); 1610, Meare (36); 1667, Mingarie (5); 1722, Mingary (3); 1723, Mingarry (3); 1737, Mingry (44); 1774, Mingarie (45); 1806, Mingary (7); 2007, Mingary (31).

Mingary Castle

Valuation: Valued as a 3½ merk land in 1610 and 1667, and a six penny land in 1723 and 1784.

Area:

A little West from Corawoolen betwixt it & the sea stands Mingary Castle with about 10 acres of arable ground around it inclosd with an old stone Dyke much decayd. The Castle seems to be in Ruins but good offices about it built by a Gentleman by the name of Campbell who resides there. Soil here of different kinds some of it light & some of it inclind to moss here is also a good deal of Limestone of which no use is made. Above the Inclosure betwixt it & the hill lies a moss some of which is draind & improvd the rest might easily be done so (Anon,1767).

In 1806 it contained 1187.19 acres, made up of 36.44 arable, 78.43 cultivated with the spade and 1072.32 acres moor & pasture. *This Farm has good Wintering, and is suited either to black Cattle or Sheep* (7).

Tenants: 1541, In the capitane of the castell of Ardmorquhen handis (55); 1690, Hugh McLaughlan and his son John (38); 1692, John Campbell and his children John, Donald, Archibald, Alexander, Colin and Katherine (38); 1716, Duncan McKenneth, Duncan McKenneth bowman, Kenneth McKenneth, Archibald Campbell, Duncan ffledger, Archibald McIlhavish, Neill McColl, John dow McArthur (37); 1722-43, held by Sir Alexander and Charles Murray of Stanhope (45); 1806, Alexander McDougal (7); 1847, 1 tenant (46).

Population: 1723, 8 families, with 13 men, 11 women and 4 children, total 28.

Settlements: NM 4954 6396. Tomachrochermean.
NM 5015 6327. Mingary.

Shielings: Not known.

Placenames:

1. **Mingary (31). 5015 6327.** ON *Mikinn Garðr* Acc. Masc. of *Mikil* 'great'. G. *Mìogharraidh,* 'the great enclosure'. Possibly G. *Mìn Gharraidh* 'the smooth enclosure'.

Mingary

National Records of Scotland, RHP72/1-8, Plan of Ardnamurchan and Sunart, Argyll, 1806.

2. **Bou Ruadh (1). NM 4783 6310.** G. *Am Bodha Ruadh* 'the red sunk rock' (from Old Norse *boða,* a breaker). Named *Sgeir Mhaol* on OS maps.

3. **Bou Carragh (1). NM 4917 6254.** G. *Bodha Carach* 'the tricky reef', or possibly *Bodha Carrach* 'stony, rocky or uneven reef'.

4. **Caisteal nan Gall (17), Castle Mingary (1). NM 5027 6312.** G. *Caisteal nan Gall* 'the castle of the strangers'. *Caisteal Mhinngaraidh* 'Mingary Castle'.

5. **Illan a Val (1). NM 5020 6300.** G. *Eilean a' Mhal* 'the island of the rent or tribute'. *Mal* s.m. 'rent or tribute'. Not a place where the estate factor would collect the rent for the district, but a dangerous skerry which could collect tribute in the form of men's lives if they were sailing along the coast with the intention of mounting a surprise attack on the castle. The OS has *Rubha a'* Mhile, with *mile* 'a mile', which is fem., but possibly not locally. Possibly a source of milestones (unlikely) or a sea marker.

6. **Skeir na Nian (1). NM 5100 6304.** G. *Sgeir nan Eun* 'the skerry of the birds' or with *nighean* 'daughters'.

7. **Tomachrochermean (3). NM 4958 6425.** G. *Tom a' Chrochaidh Mheadhain* 'Middle Gallows Hill'. Named *Tom a'Chrochaidh* on OS maps. The settlement here was for cottars, some of whom paid no rent. In 1854, there were 7 cottar families and a population of 36.

8. **Glac Leum (34). NM 4837 6294.** G. *A' Ghlac Leum* 'the leap-over defile', which sounds plausible, but it may be **Glaslean** G. *Glas-Eilean* 'the blue/grey island'.

9. **Doirlinn (34). NM 4837 6294.** G. *An Dòirlinn* 'the (tidal) isthmus'.

10. **Caim (2). NM 5117 6460.** G. *A' Chaim* 'the bend'. The dwelling here is called '**Camphouse**', but Henderson c. 1910, states that this name was not used locally.

OCKLE : Ocal : field

1541, Ockill (55); 1610, Ockill (36); 1618, Ochill, Okill (8); 1667, Orkill (16); 1722, Ockle (3); 1723, Okill (3); 1784, Ochkill (45); 1806, Okill (7); 1828, Ochkill (6); 2007, Ockle (31).

Ockle

Valuation: Valued as a 2½ merk land in 1541 and 1610, and part of the 8 penny lands of Swardlehiel in 1723. A five penny land in 1784 with Suaradilcheille a three penny land (45).

Area: In 1806 it contained 3435.17 acres, made up of 59.19 acres arable, 92.02 cultivated with the spade and 3284.05 acres of moor and pasture (7).

Tenants: 1541, Farquhar McAlister McRanald (55); 1618, Allan McAllan VcPersone (8); 1732, let with Swardlecheil to John McLachlan, Archibald McEacharn, Duncan McLachlan, Donald McKenzie, John Stewart and Donald Cameron; 1806, Allan Cameron and small Tenants. (7); 1828-1840, 2 tenants (6); 1847, 7 tenants (46).

Population: 1841, Eligidle had 3 occupied houses, with 15 people.

Settlements: NM 5558 7042. Ockle.
NM 5747 7130. Eilagadale.
NM 5603 6971.

Shielings: NM 547 661. Shieling and bloomery.
NM 5602 6972. 'Allt Eas a' Ghaidheil'.

Placenames:

1. **Ockle (31). NM 5558 7042.** G. *Ochcall* or *Ocal,* which Iain Tàilleir takes to be an ON loan-name containing *vôllr* 'field'. Local tradition suggests that it is derived from G. *ochd* 'eight', the burn being the eighth on the path to (or from ?) Gorten.

2. **os Call (1). NM 5744 6974.** Alasdair Cameron equates this name with the *Allt Eas a' Ghaidheil* 'the waterfall of the Gael's burn'. Could 'os Call' be *eas* 'waterfall' (of the) *call* 'calamity' ?

3. **Maolalanaranach (1). NM 5739 7004.** G. Possibly *Maol Ailein Aranaich* 'the hill of Alan *Aranach* (from Arran').

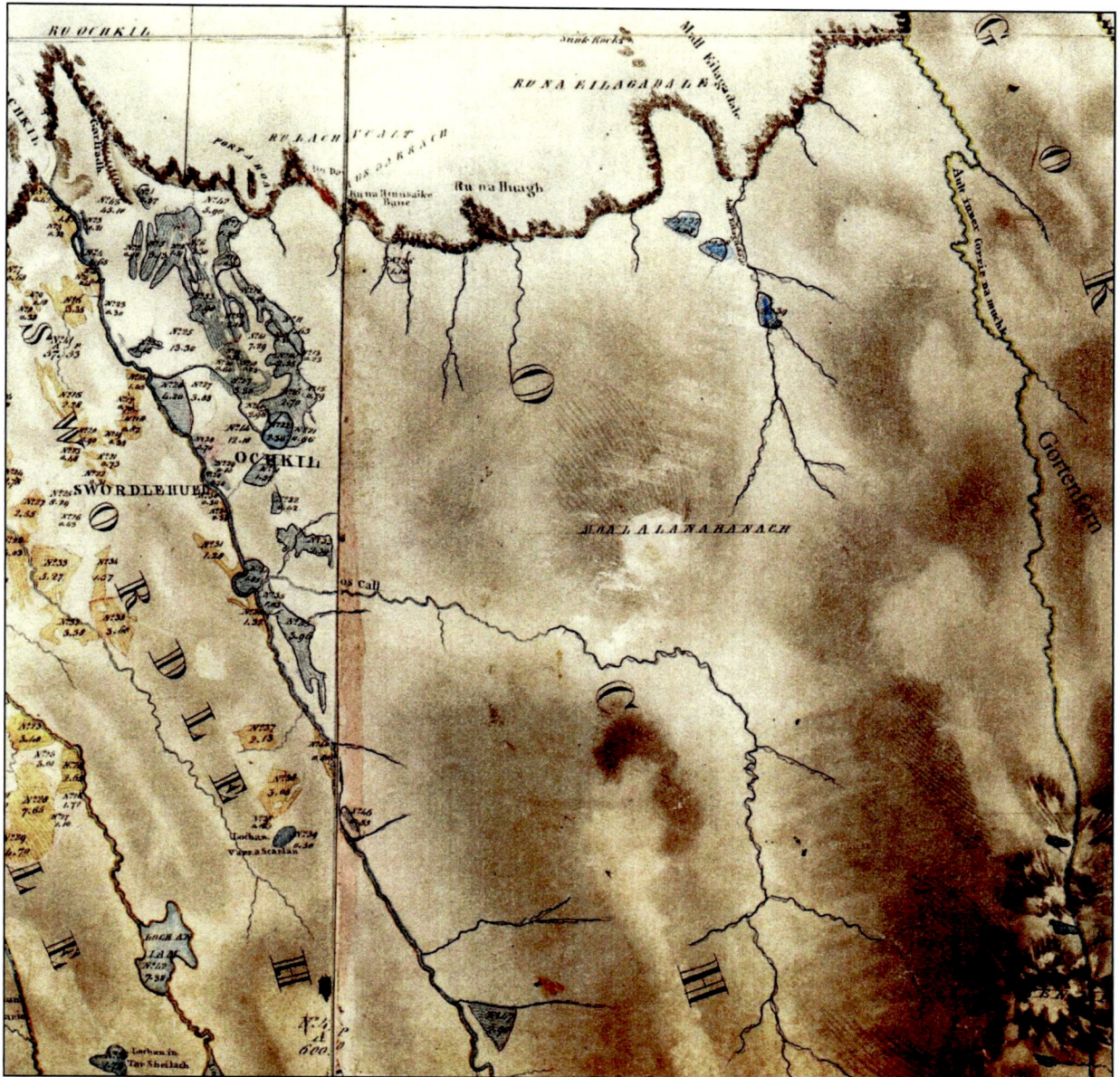

Ochkil (detail) National Records of Scotland, RHP72/1-8, Plan of Ardnamurchan and Sunart, Argyll, 1806.

4.   **Ru na Hinnsaike Bane (1). NM 5610 7142.** G. *Rubha na h-Innseige Bàine* 'the point of the little white field'.

5.   **Camus Darrach (1). NM 5600 7130.** G. *An Camas Daraich* 'oak bay'.

6.   **Ru Lach a' Vualt (1). NM 5508 7142.** G? *Rubha Lag a' Mhuilt* 'the headland of the hollow of the wedders'. *Mult* (gen.) *muilt* m 'wedder sheep ie. castrated tups'.

7.   **Port a' Hoat (1). NM 5558 7125.** G. *Port a' Choit* 'the port of the small boat'. *Coite* 'a small fishing boat, coracle'. This port was probably only suitable for a small boat, but there are three 'ports' here and it is uncertain which one this name refers to.

8.   **Garliadh (1). NM 5508 7142.** G. *An Gàrradh Liath* 'the grey dike'. *Gàrradh* 'a wall, dyke'. A line of cliffs on the east side of Ockle Point.

9.   **Camas Ochkil (1). NM 5474 7155.** G. *Camas Ocail* 'Ockle Bay'.

130

10. **Mull Eilagadale (15).** NM 5732 7176. G. *Mull Eilgeadail* 'the cape or jutting rock of Eilagadale'. From Old Norse *mulr* 'snout or jutting rock'. The name probably refers to the eastern promontory of *Rubha Eilagadale*.

11. **Elgadale (2).** NM 5747 7130. G?. *Eilgeadail.* The name ends in ON *dalr* 'valley', but the first element is obscure. Probably not ON *elg* 'noble' or *elgr* 'elk'. G. *eileag* 'a deer trap' has been suggested.

12. **Coire Cullah (47).** NM 5604 6975. G. *Coire Caol* 'the narrow corrie', or possibly G. *culadh* '

13. **Port an Drian (47).** NM 5476 7143. G. *Port an Droighionn* 'the boat landing of the blackthorn'.

14. **Corrynnenach (55).** NM 5747 7130. G. *Coire nan Each* 'the corrie of the horses'. The settlement of Corrynnenach is not known but it occurs between Ochchall and Clais, and is here assumed to be an alternative name for *Eilagadail.*

Settlement where the Allt Eas a' Ghaidheil meets the Allt Ockle

Ockle from Dùn Mòr, Swardle

Port a' Choit: the house, the sweep of the bay and one of the three ports.

The dùn on Rubha na h-Uamha (but where is the cave?)

Ochkil sheep-fank

Ru Lach a' Vualt and Ru Ardrimonish from Garliabh

ORMSAIGBEG : Ormasaig Beag : the lesser (farm of) Orm's bay

1541, Ormisagebeg (55); 1610, Ormissabege (36); 1667, Ormsagbeg (5); 1722, Ormsaigbeg (3); 1723, Ormasaigbeg (3); 1737, Ormsagbeg (9); 1784, Ormsagbeg (45); 1806, Ormsabig, Ormsegbeg (7); 2007, Ormsaigbeg (31).

Ormsaigbeg and Kilchoan Bay

Valuation: Valued as a 20/- land in 1610 and a five penny land in 1723. In 1541, *Thir ar the landis underwrittin that is allegit to pertene to the kirk, Ane mark land of the saidis landis of Ormissagbeg fest to Ycomkle (Iona Abbey) and ane hale mark land therof to Sant Peteris Chapell (55).* St. Peter's Chapel has not been identified, but it could be Restenneth Priory, a house of the Augustinian Canons Regular.

Area: In 1806 it contained 1382.13 acres, made up of 19.91 acres arable, 57.18 cultivated with the spade and 1305.04 acres of moor & pasture.

Tenants: 1541, Angus McAlister McKane (55); 1693, Donald McDonald and his wife Mary McDonald (38); 1716, Donald Mcffinlay VcEan Vuy, John Mcffinlay VcEan Vuy, ffinlay McGreisich, Donald McGlashan, Gilmichael McIlmichael, Hendry McKendrick, Patrick McKellar (37); 1737, Neil Campbell, (½ of the township), Alexander McDonald (⅛), Donald McIllvraw (⅛), Hugh Henderson (⅛) and Kenneth McKenzie (⅛) (9); 1806, Alexander McDougal, who also farmed Mingary (7); 1828, 20 crofters on a year to year tenancy (6); 1864, 1 farmer and 3 crofters were 'cleared' (42).

Population: 1723, 6 families with 8 men, 10 women and 10 children, total 28.

Settlements: NM 475 635. Ormsaigbeg. Fifteen structures shown on the estate plan.

Shielings: NM 4360 6440. *Reidh-dhail*. A developed shieling with an extensive field system and at least six low drystone buildings.

Placenames:

1. **Ormsaigbeg (31). NM 475 635.** G. *Ormasaig Beag* 'the lesser (farm of) Ormsaig'. ON *Orm* 'a snake', or possibly a Norse personal name, *Orm*. G. *aig* from ON *vik* 'a bay'. G. *beag* 'little'. A loan-name likely to be from ON *Ormsvík,* 'Orm's bay' or 'the bay of the snake' (or possibly a sea-serpent)'.

2. **Black Lochs (1). NM 4380 6478.** G. *Na Lochan Dubha.*

3. **Redel (1). NM 4375 6445.** G. *An Rèidh Dhail* 'the smooth field'.

4. **Loch Abraid Buidh (1). NM 4465 6415.** G. *Loch a' Bhràghaid Bhuidhe* 'the loch of the yellow throat'.

Ormsaigbeg

National Records of Scotland. RHP72, Plan of Ardnamurchan and Sunart, Argyll, 1806..

An Rèidh-dhail and *Na Lochan Dubha*

5. **Ru Aoin a' Chaol (1). NM 4510 6262**. G? Named *Rubhan a' Chall* on Ordnances Survey maps. *Rubhan* 'small headland'; *Call* 'loss, damage, calamity'. 'the small point of the calamity'.

6. **Bun an Uilt Gairbh (1). NM 4556 6278**. G. *Bun an Uillt Ghairbh* 'the foot of the rough burn'.

7. **Croch na Cnoden (1). NM 4592 6275**. ?G. *Cnoc nan Cnòdan* 'the hillock of the gamets, gurnets or gurnard' (*Trigla gurmadus),* a species of fish.

8. **Coidhluim (1). NM 4654 6222**. G? The OS maps places this name at the top of the cliffs of *Sròn Beag* and names it *Cuingleum*, while the Estate Plan shows it as the point jutting into the sea. Possibly *coinglidh* 'narrow defile'.

9. **Scridens Ruoy (1). NM 4592 6247**. G. *Sgrìodan Ruadh?* 'the red scree'. Old Norse *skridda* 'landslip'.

10. **Bou na Keill (1). NM 4780 6300**. G. *Bodha na Cille* 'the skerry of the church', 'the church reef'.

11. **Uilt Gairbh (1). NM 4548 6300**. G. *An t-Allt Garbh,* unless it is the plural *Na h-Uillt Gharbha* 'the rough burn(s)'. Derived from No.5, *Bun an Uilt Gairbh.*

Caisteal Dubh nan Cliar looking towards Ben Hiant.

12. **Castle Dou na Cliabh (1). NM 4731 6312**. G. *Caisteal Dubh na Cliabh* 'the black castle of the chest, wickerwork straight jacket or cheese kist'. Named *Caisteal Dubh nan Cliar* 'the black castle of the minstrals', on OS maps. The castle is at least 3m longer than shown on the RCAHMS plan, making the doorway with the rebated 'entrance' doorway and bar-hole an internal feature, and suggesting that the rock shelter might have been a prison (for a man in a straight jacket ?). There is a possibility that the area adjacent to the castle was used as a boatyard and noost for overwintering the MacIain of Ardnamurchan fleet during the Mediaeval period.

13. **Sgeir MacArailt (34). NM 4735 6294**. G. *Sgeir MacArailt* 'MacRaild's Skerry'.

14. **Garbh Allt (34). NM 4731 6285**. G. *An Garbh Allt* 'the rough burn'.

15. **Lochan an t-Sàlann (34). NM 4782 6354.** G. *Lochan an t-Salainn* 'the lochan of the salt'. A small, dirty pool on the salt marsh fore-shore. There was a jetty here for the Mull steamer and it was used by the local fishermen.

16. **Croit a' Gobhain (34). NM 4720 6286.** G. *Croit a' Ghobhainn* 'the blacksmith's croft'. The current name for the house and croft is *Cruachan* 'a conical hill or mountain top'.

17. **Bodha Cùil Ard (34). NM 4720 6271.** G. *Bodha Cùl na h- Àirde* 'the sunk rock at the back of the point', but it might be for Gaelic *Bodha a' Chùil Àird* '...of the high back end'. Named *Bogha Caol Ard* 'the high narrow skerry' on OS maps.

18. **Cùil Odhar (34). NM 4720 6286.** G. *An Cùl Odhar* 'the dun coloured narrows'.

19. **Creag Ard (34). NM 4755 6240.** G. *A' Chreag Àrd* 'the high crag'. Also known as *Mull Dubh* 'the black promontory'. Described as being 'Catie Connel's Croft'.

20. **Rubha Chaisteal Dubh (34). NM 4742 6203.** G. *Rubha a' Chaisteil D(h)uibh* 'the point of the black castle'.

21. **Glac a Chleirich (17). NM 47 63.** G. *Glac a' Chléirich* 'the hollow of the clergyman (or the clerk)'.

22. **Cnoc na Gour (1). NM 4490 6330.** G. *Cnoc na Gobhar* 'the hill of the goat'.

23. **Beinn nan Cathan, 742 ft (15). NM 4420 6340.** G. *Beinn nan Cathan* 'the hill of the barnacle geese'. Named *Beinn nan Codhan, 227m* on OS maps. *Codhan/Cobhan* 'hollow, small creek, box or coffin'.

An Acairseid

Ormsaig Mòr and Ormsaig Beg from Tom na Cille

ORMSAIGMORE : Ormasaig Mòr : the greater (farm of) Orm's Bay

1541, Ormissagemoir (55); 1610, Ormissagemoir (36); 1618, Ormesag, Ormessag (8); 1667, Ormsagmoir (5); 1722, Ormsaigmore (3); 1723, Ormasaigmore (3); 1739, Ormsaigmore (9); 1784, Ormsagmore (45); 1806, Ormseg more (7); 2007, Ormsaigmore (31).

Ormsaigmore from the SW

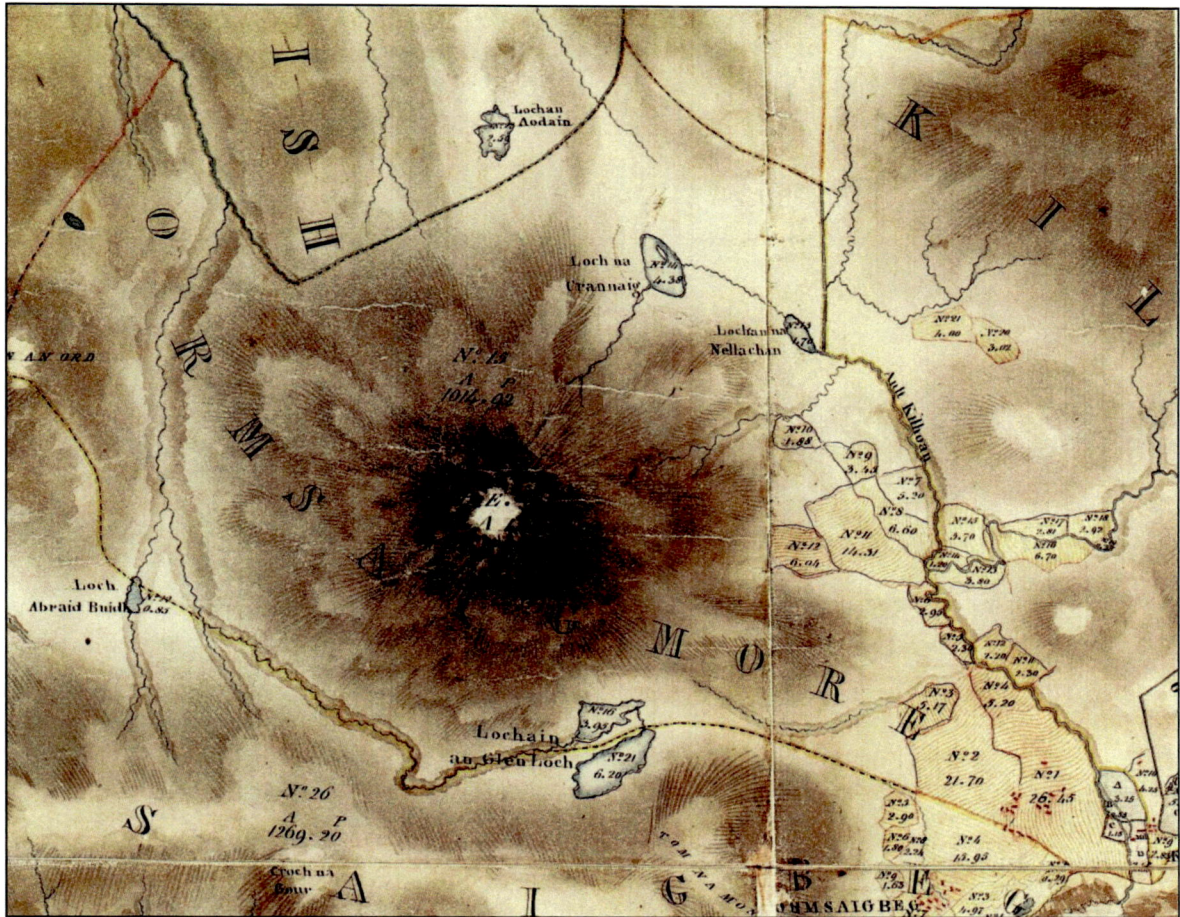

Ormsaigmore

National Records of Scotland, RHP72/1-8, Plan of Ardnamurchan and Sunart, Argyll, 1806.

Valuation: Valued as a 2½ merk land in 1610, a four penny land in 1723 and a five penny land in 1784 (45).

Area: In 1807, the 1126.20 acres consisted of 62.58 ploughable land, 33.47 cultivated with the spade, 5.20 pasture, 10.03 lochs and 1014 moorland.

Population: 1723, 6 families, with 9 men, 8 women and 12 children, total 29.

Tenants: 1541, Angus McAlister McKane (55); 1618, Donald McEane, with Johnne, Angus Allester, Donald and Archibald McEane his sons (8); 1698, Donald McLauchlan, his wife Anna NcHendry and their children, John, Donald, Malcolm, Mary and Christian (38); 1716, John McIlleriach, piper, Hector Mcphaill, Alexander Mcphaill, Malcolm Mcphaill, Donald Campbell, Donald McKendrig and Donald McIlleriach (37); 1737, Duncan McKenzie (½ of the township), Arch'd Murray (¼) and Janet Campbell, relict of John Campbell (¼) (9); 1806, Duncan McPhail and four others; 1828, 8 tenants on a year to year let (6); 1852, 6 tenants were evicted for non-payment of rent (53); 1864, 4 farms were 'cleared' (42).

Settlements: NM 476 638. Ormsaigmor. Thirteen structures shown on the estate plan of 1807.

Shielings: Not recorded.

Placenames:
1. **Ormsaigmore (31). NM 476 638.** G. *Ormasaig Mòr* 'the greater farm of Ormsaig'. The Gaelic word *Ormasaig* is likely to be from ON *Ormsvík* 'the bay of the snake' or from the Norse personal name *Orm.* See Ormsaig Beag above.

2. **Leacann (34). NM 4695 6433.** G. *An Leacann* 'the rocky slope'. *Leac* 'bedrock or flat stones'.

138

3. **Lochan na Nellachan (1). NM 4700 6555.** G. *Lochan nan Ealachan* 'the lochan of the mute swans'. *Eala* 'mute swan', *ealachan* s. f.

Greadal Fhinn, a Neolithic Chambered Cairn

Greadal Fhinn looking NW

SKINAD : Sgìnid :

1541, Skeneith (55); 1610, Skeneith (36); 1667, Skenith (5); 1722, Skyneid (3); 1723, Skinead (3); 1724, Skenning (47); 1737, Skinad (9); 1741, Skinnet (12); 1767, Skinned (29); 1784, Skinid (45); 1806, Skinid (7), 1828, Skinad (6).

Skinad

Valuation: Valued as a 2½ merk land in 1541 and in 1610, and a five penny land in 1723.

Area: *Above Mingary Moss in the Hill lies a large Town calld Skinned with about 30 acres of arable ground round it. Soil of different kinds this might be enlarged and improved if the lime was used about 10 acres of this is inclosed by a Feal Dyke. The Houses all built with stone & Feal & in good Repair, at the back of which is a large hill pasture* (Anon, 1767).

In 1806 it contained 1133.40 acres made up of 85.31 arable, 15.20 cultivated with the spade and 1032.89 acres moor and pasture. *This Farm contains more arable land by the plough than most upon the Estate but the soil is thin , though it is under a pretty good crop this year. It is also oppressed with a number of Tenants (7).*

Tenants: 1541, Donald McAlister McKane (55); 1737, Duncan Stewart (9); 1741, Angus McDonell, tenent (12); 1806, John McLauchlan and ten others (7); 1828, John McColl on a year to year let (6).

Population: 1723, 9 families, with 12 men, 14 women and 20 children; total 46.

Settlements: NM 5165 6436. Skinad.

Shielings: Not recorded.

Placenames:

> **1. Skinid (2). NM 5165 6436. ?G. *Sgìnid.*** This is an early and enigmatic name. Professor W.F.H. Nicolaisen has suggested that Skinnet in the valley of the River Thurso, Caithness is derived from ON *skínandi* 'the shining one', which relates it to the river (50), but this does not apply to Skinnet at the head of the Kyle of Tongue, and neither is it appropriate here.

Skinad (detail)   National Records of Scotland, RHP72/1-8, Plan of Ardnamurchan and Sunart, Argyll, 1806.

2. Ballach man a Ghaoil (1). NM 5000 6700. G. *Bealach Màm Ghaothail* 'the pass of the windy hill'. *man* or *màm'* a hill of particular form, slowly rising and not pointed'; 'a large round hill'.

3. Ault Raevaulone (1). NM 5117 6460. G. *Allt Ruighe Mhuilinn* 'the burn of the shieling of the mill'. Named *Allt Coire Mhuilinn* 'the burn of the corrie of the mill' on OS maps.

SWARDLE CORRACH: Suardal C(h)orrach : Steep Swardle.

1541, Swardillcorroch (55); 1610, Suardillcorroch, Swardilcorroch (36); 1618, Surdill (8); 1667, Swardillcorroch (5); 1722, Suardilehoreach (3); 1723, Suardile horrich (3); 1737, Swardal chorrich (9); 1784, Suaradlcorach (45); 1806, Swardleharach (7); 1807, Swordlechorrach (1); 2007, Swardle (31).

One of several creel buildings on Swardle Corrach

Valuation: Valued as a 2½ merk land in 1610 and a 5 penny land in 1723.

Area: The 479.49 acres were made up of 38.80 arable, 27.61 cultivated with the spade and 413.08 moor & pasture.

Tenants: 1541, Johanni (Johne Erenoch ?) (55); 1618, Donald Roy McTaves [or McAngus] and Rorye McGillimichaell (8); 1737, Archibald McNiven, along with Kilmory (9); 1739, Donald McLauchlane (13); 1806, Duncan McColl and 5 others; 1828-1840, 6 tenants (6).

Population: 1723, 6 families, with 9 men, 9 women and 6 children; total 24.

Settlements: NM 5420 7054. Swardle chorrich.

Shielings: Not known.

Placenames:

1. **Swardal Chorrich (9). NM 5424 7038.** G. *Suardal C(h)orrach* 'Steep Swardale'. *Corrach* (adj) 'steep, unstable'. Possibly *Suardal Choireach* with gen. of *Coireach* 'full of circular hollows'. Swardal probably contains a loan-name from ON *Svarð-dalr* 'sward/grass-valley'.

2. **Ru na Aird (1). NM 5384 7188.** G. *Rubha na h-Àirde* 'the promontory of the height'. Named *Garbh Rubha* 'the rough promontory' on OS maps.

3. **Lochan Horarie (1). NM 5503 6820.** G. *Lochan* h-...? Named *Lochan na Corra* or *Lochan Chorra* 'the lochan of the heron' on OS maps. Horarie may represent a Gaelic form of an ON loan-name ending in *ærgi* 'shieling'.

4. **Hammas (47). NM 5365 7064.** G. *An Camas* 'the bay'.

5. **Uamh Tuille (47). NM 5336 7085.** G. *Uamh Tuille* 'the cave of the hole'. A wet cave with a hole in the back, which gives the observer a concealed view across the bay. *Uamha Thuill* appears on the OS maps at NM 5379 7105. This consists of two rather open caves on the shore with a holed rock wall between, and a pool of water in the floor where St. Columba is said to have baptised two robbers.

6. **Rubha na Cisteachan (47). NM 5356 7080.** G. *Rubha na Cisteachan* 'the headland of the chest/coffins'.

SWARDLE HUEL: Suardal Cheil : Hidden or Sheltered Swardale.

1541, Suerdllbeg (55); 1610, Suerdillbeg (36); 1667, Swardilbeg (5); 1686, Surrodellcheill (10); 1667, Swardilbeg (5); 1722, Suardileheile (3); 1723, Suardill Chastill (3); 1723, Swardilcheul (4); 1734, Sordaalchail (45); 1737, Swardalchail (9); Sorradale chuerlie (13); 1741, Suadalechoile, (12); 1784, Suaradilcheill (45); 1806, Swordlehuel (7); 1828, Swardlechuel (6); 2007, Swardle (31).

Valuation: A 2½ merk land in 1610. Valued as an 8 penny land in 1723 & 1732, and 7 pennies in 1737, but this valuation probably included Ockle, a five penny land, with Suaradilcheille a three penny land in 1784 (45).

Area: Of the 767.87 acres, 41.94 were arable, 13.97 cultivated with the spade, 668.71 moor & pasture and 43.25 lochs. *This is a very good grazing, but if occupied as a grazing, it would afford Bread only for one Family. If these three Swordles were joined into one Farm, they would make a very good one – Very fit for the keep of Cheviot Sheep.*

Tenants: 1541, Donald McAlister McKane (55); 1686, Duncan VcIan Vain McDonald (19); 1693, Angus McIllvrae and his wife Mary NcPherson (38); 1716, Duncan Stewart, Duncan McIlleise, his servant, Donald McColl his tenant, John McCraiggan, his servant, Duncan McIllom, his workman, Angus Cameron, workman, John McArtne, John McGlashen, John McKenneth, Alexander Cameron, John McKenzie (37); In 1732, let for 5 years, to John McLachlan in Branallt (1½d), Archibald McEacharn there (1d), Duncan McLachlan there (1d), Donald McKenzie in Swardle cheil ((1½d), John Stewart in Achatenie (1½d) and Donald Cameron in Glenborradale (1½d) (14); 1734, Angus McDonald, Archibald McEacharn, Donald Cameron, Duncan McLachlan, John McLachlan, Angus and John Blacks (45); 1737, Angus

143

The Swardle Townships (detail)   National Records of Scotland, RHP72/1-8, Plan of Ardnamurchan and Sunart, Argyll, 1806.

McDonald (3d), Arch'd McEacharn (1d), Donald Cameron (1½d), Duncan McLauchlan (1d) and Angus Black (½d); 1740, Donald Cameron McAllan Vic alister, residenter (12); 1806, Donald McColl and 4 others; 1828-1840, 6 tenants (6); 1847, 6 tenants (46). 1852, 12 tenants were evicted for non-payment of rent (53).

Population: 1723, 9 families with 14 men, 12 women and 11 children, total 37.

Settlements: NM 9457 7017. Swardlehuel.

Shielings: NM 558 694.

A recent house with possibly earlier ones marked in red

Excavating the Viking Boat Burial

Placenames:

1. **Swardlehuel (7). NM 5494 7008.** G. *Suardal Cheil* 'hidden or sheltered Swardle', or G. *Suardal Chaol* 'Narrow Swardle'.

2. **Lochan Shien (1). NM 5711 6683.** G. *Lochan an t-Sìdhein* 'the lochan of the fairy knoll'. *Sìthean (*gen. *an t-Sìthein)* m 'fairy knoll'. Named *Lochan nan Sioman* on OS maps. *Sìoman (*genitive *sìomain)* m 'a straw rope'.

3. **Lochan na Three Criagh (1). NM 5682 6643.** G. *Lochan nan Trì Crìoch* 'the lochan where three boundaries meet'. *Tri* 'three'. *Crìoch* ' march, boundary'. Named *Lochan na Carraige* 'the lochan of the rock' on OS maps. *Carraig (*gen. *carraige)* f 'a rock'.

4. **Illan Swardle Beg (1). NM 5468 7147.** G. *Eilean Suardail Beag* 'the little island of Swardle' or 'the island of Little Swardle'.

5. **Swardilbeg (5). NM 5494 7008.** G. *Suardal Beag* 'Little (lesser) Swardle'. An alternative name for Swardle Cheil.

145

6. **Lochan Varr a Scarlan (1). NM 5600 6895**. ?G. *Lochan Bhàrr an Sgàrlain,* but the meaning is unclear. Named *Lochan na h-Eilde* by the OS.

7. **Port Drain (1). NM 5458 7100**. G. Possibly *Port Dreathain* 'the port of the wren'. This is the boat landing at Inverochkil.

8. **Loch An Iam (1). NM 5560 6850**. Named *Lochan an Ime* 'the lochan of the butter' ie rich grazing, on OS maps.

9. **Cnoc na Ceapach (47). NM 5517 7035**. G. *Cnoc na Ceapach* 'the hill of the tree trunks' or *Cnoc na Ceapaich* s.m. 'the hill of the tillage plot'.

10. **Cladh Aindreis (28). NM 547 707**. G. *Cladh Aindreis* 'Andrew's grave or burial ground'. *Aindreis* is an unusual name in the area, but in the Argyll rental of 1506 it is recorded that 'Sunart in Morvern' was leased to *"domino Andree McCachin, rectori de Ardmurchy",* and Andrew MacEachern, undoubtedly the same person, who died some time before 1515, was parson of *"Ellenenan et Kilquhoan in Ardnamurchane"* (56). Andrew was probably a MacEachern of Killellan in South Kintyre, and not of the *Siol Eachern* MacLeans of Kingairloch, who produced a number of prominent ecclesiatics in the late mediaeval period (55, 56). The name is now applied to the Neolithic chambered cairn by the burnside.

11. **Lochan Smear (1). NM 5591 6800**. ? G. *Lochan nan Smeur* 'the lochan of the blackberrys'. Named *Lochan nan Dearcag* 'the lochan of the little berries' on OS maps. This is more likely to refer to blaeberries than to brambles.

Excavating the sword in the Viking Boat Burial, 2011

SWARDLE MORE: Suardal Mòr : the greater (farm of) Swardle

1541, Suerdillmoir (55); 1610, Suerdillmoir (36); 1722, Suardilmore (3); 1723, Suardill More (3); 1723, Swardilmoir (7); 1737, Suardilmore (9); 1784, Suaradilmoire (45); 1828, Swardlemor (6); 2007, Swardle (31).

Swardle More from Swardle Cheul

Excavating the Viking boat burial on Swardle Cheil, looking across to Swardle More

Valuation: Valued as a 2½ merk land in 1610 and a 5 penny land in 1723 (7).

147

Area: In 1807 the 471.83 acres were made up of 68.49 ploughable land, 7.71 cultivated with the spade, 10.47 loch, 60.16 pasture and 325.00 acres of moorland.

Tenants: 1541, Donald McAlister McKane (55); 1716, Allan Stewart, his sons Alexander and James, his brother Duncan, Alexander Stewart, Donald McIleise and his son, also Donald, Patrick McIlvra, Dugald Mcpherson, Angus McIlvra and John oig McEachern (37); 1737, Archibald McDonald (9); 1739, Donald McDonald, Ewan Black and John McColl (13); 1828-1840, 6 tenants (6); 1847, 6 tenants (46). 1852, 6 tenants were evicted for non-payment of rent (53).

Population: In 1723, 5 families, with 10 men, 6 women and 8 children, total 24.

Settlements: NM 5470 7033. Swardle More.

Shielings: NM 553 687.

Placenames:

1. **Suardilmore (9). NM 5470 7030.** G. *Suardal Mòr* 'the greater (farm of) Swardle'.

2. **Lochan Derry na Tua (1). NM 5593 6744.** G. *Lochan Doire na Tuatha* 'the lochan of the grove of the tenantry'. *Doire* (gen *doire*) m/f. 'a thicket, small wood'. *Tuath* m. 'tenantry, peasantry, country people', but possibly with *Doire Tuath* 'northern grove'. Named *Lochan na Tuaidh* on OS maps.

Camas Suardal Mòr with *Port an Eilean Mòr* at its head.

3. **Illan More (1). NM 5450 7106.** G. An t-*Eilean Mòr* 'the big island'.

4. **Maoll Keir (1). NM 5437 7887.** This is an odd name as it is a low tidal rock, and *sgeir* would be more appropriate here. Possibly with *maol* 'bare, bald, blunt, barren' and *sgeir* 'skerry'.

5. **Skeir na Scarrabh (1). NM 5435 7100.** G. *Sgeir nan Sgarbh* 'the skerry of the cormorants'. ON. *'Skarfr sker'*

6. **Lochan in Tar Sheilach (1). NM 5555 6811.** G. *Lochan an Tòrr Sheilich* 'the lochan of the willowy hill'. Named *Lochan an t-Seilich* 'the lochan of the willows' on OS maps.

7. **Parc a' Chladaich (47). NM 5450 7083.** G. *Pairc a' Chladaich* 'the park by the shore'.

TORNAMONY : Tòrr na Mòine : the hill of the peat bog

1541, Tornomonoch (55); 1610, Tornomonoch (36); 1667, Tornamont (5); Tornamona, (3); 1723, Tornamonah & Derrybuie (3);1723, Tornamoany (4); 1737, Tornamoan, (9); 1784, Turnamoen (45); 1807, Tornamony (29); 1827, Tornamoany (6); 2007, Torr na Moine (31).

Settlement by the *Allt Torr na Moinne*

Later settlement, a shepherd's house and byre

Valuation: Valued as a 2½ merk land in1610 and a 5 penny land in1723.

Area: In 1806 it contained 745.68 acres, made up of 15.69 arable, 75.71 cultivated with the spade, and 654.28 acres of moor and pasture.

On a little rising W from Camisingall lis Toutnamon a small place with about 5 acres of arable ground but very extensive pasture & Glens above it (29).

Tornamony   National Records of Scotland, RHP72/1-8, Plan of Ardnamurchan and Sunart, Argyll, 1806.

Tenants: 1716, Duncan McCallman, John dow Mcphaill, John Mcphaill, a boy, Donald McKenzie, Donald McLea, servant, John McIlonoch, John McInnish Smith, a vagabond (37); 1737, Duncan Stewart (9); 1807, John McColl and twelve others (7); 1828-1833, John McColl (6).

Population: 1723, 5 families, with 9 men, 6 women and 4 children; total 19.

The promontory fort

From the *port* on Ardslignish

Shepherd's house

Shepherd's house and byre

Settlements: NM 5458 6547. Enclosure and structures.

NM 5570 6224. Tornamony. Numerous buildings demolished to build the fank.

NM 5470 6515. *Doire Buidhe*. A number of buildings at the foot of the ravine.

Smiddy, location not known.

Shielings: NM 5495 6444

Placenames:

1. **Tomamona (3). NM 5570 6224.** G. *Tòrr na Mòine* 'the hill of the peat bog'.

2. **Chnab na Ceanamhor (1). NM 5520 6157.** ? G. *Cnap na Ceannamhoir* 'the hillock of the great headland'.

Sgoim from Ardslignish, looking across Camas nan Geall.

3. **Sgoim (1). NM 5525 6143**. G. *An Sgaoim* 'wandering', or 'sudden terror/ starting from fear or terror'. *Sgaoim* f 'fright, alarm'. *Sgoim* f 'wandering about'. A skerry that frightened people by appearing out of nowhere. It is marked on the 1ˢᵗ Edition OS maps but omitted from more recent surveys. Both English translations are appropriate for a skerry that is not always visible.

4. **Craig Chorrach (1). NM 5534 6163**. G.?. *A' Chreag Chorrach* 'the rock of the sheep'. *Caora* gen. *caorach*, pl. *caoraich,* gen. pl *caorach* f. 'sheep'. Possibly G. *Creag Chorrach* 'the steep rock'.

5. **Derrybuie (3). NM 5470 6515.** G. *Doire Buidhe* 'the yellow grove'. A pre-1723 settlement.

Doire Buidhe and Loch Mudle

153

The hill fort, with Ardslignish beyond

Tornamony

REFERENCES

1. NRS: RHP 72. William Bald's *Estate Plan of the Barony of Ardnamurchan and Sunart, 1806/7.*
2. Henderson, Angus, (1916). 'Ardnamurchan Place-names' in *The Celtic Review* **10** 1914-16 pp.149-168.
3. NLS: Adv. Ms. 29/1/1 Vol.VII, f.139-43. Rental of Ardnamurchan, 1723.
4. Murray of Stanhope, Sir Alexander, 1740. *The True Interests of Great Britain, Ireland and Our Plantations.* Appendix entitled 'Anatomie of the Parish and Barony of Ardnamoruchan and Swinard, 1723'.
5. Register of the Great Seal of Scotland (RMS) Vol. XI No. 1105.
6. NRS: AF49/3. Report on the State of Farms on the Barony of Ardnamurchan and Sunart. Thomas Anderson, 1829.
7. NRS: AF49/2/1. Valuation of the Estate of Ardnamurchan. Alex. Low, 1807.
8. Register of the Privy Council Vol. XII. A number of McEanes put to the horn for besieging Mingary Castle, 1622.
9. NRS: CC2/2/30/2. Summons by the Duke of Argyll and Crown vs Parishioners in Ardnamurchan, 1734.
10. Francis J Grant Ed (1902). *Commissariot Record of Argyle; Register of Testaments, 1674-1800.* Scottish Record Society, Edinburgh.
11. NRS: GD241/166. Thomson and Baxter WS Ms.
12. NRS: GD202/46. Dunstaffnage Papers. Theft of stock from Achosnich and Skinnet, 1734-41.
13. NRS: CC2/2/33/5. Summons by John Richardson, Factor to sundry tenants in Ardnamurchan, 1739.
14. Tack dated 5th January 1732 between Sir Alexander Murray of Stanhope and McLauchlan & oyds.
15. NRS: RHP 3345. Sale Plan of the Estate of Mingary, 1885, but based on RHP 72.
16. RMS Vol. XII No. 1195.
17. NLS: Ms.369 Place-names of Ardnamurchan, Sunart etc. Probably by Charles M. Robertson, c. 1900.
18. Admiralty Chart No.3185, *Loch Sunart*, 1864. Surveyed by Capt. E J Bedford R.N.
19. Donald Campbell in '*Tocher* **28**. (1978) pp.198-9. School of Scottish Studies, University of Edinburgh.
20. NLS: Adv.Ms 29/1/1 Vol VII, f.133-6. Rental of the Barony of Ardnamurchan and Sunart, 1722.
21. MacMillan, Catriona and Hodgekinson, Ursula et al (1981). *Ardnamurchan : Annals of the Parish.* Metro Press, Edinburgh.
22. National Library of Scotland Ms.473. Exercise book with local stories, place-names and Gaelic.
23. CL/A9/3 Langland's Map of Argyll, 1801.
24. NRS: GD241 Box 164, 2,3. Alexander Bruce's *Plan of Loch Sunart* prepared on behalf of Sir Alexander Murray of Stanhope and circulated in 1733. Also appears as RHP 9103.
25. SC 54/2/17/65/6. Inverary Sheriff Court Records. Summons of MacOlony Camerons and others for theft of lead from Mingary Castle in July 1741 by Campbell of Lochnell.
26. Cameron, Alasdair 'North Argyll' (1954). *Loch Sunartside Memories,* Oban Times.
27. Imrie, John Editor (1969). *Justiciary Records for Argyll and the Isles.* 1664-1742.144. Stair Society. Vol. 25.
28. RCAHMS *Argyll; An Inventory of the Monuments, Vol. 3. Mull, Tiree, Coll & Northern Argyll.* Royal Commission on the Ancient and Historical Monuments of Scotland, 1980.
29. Estate description on the eve of sale, 1767. Anon.
30. Maclean, Alasdair (2001*). Night Falls on Ardnamurchan. The Twilight of a Crofting Family.* Birlinn.
31. Ordnance Survey Explorer Map, Sheet 390, *Ardnamurchan, Moidart, Sunart and Loch Shiel.* 1: 25,000 Scale, 2007.
32. Valuation Roll of Argyll, 1873.
33. Black, Ronald (1986). *Mac Mhaistir Alasdair; The Ardnamurchan Years.* Society for West Highland and Island Historical Research.
34. Archie Campbell. School of Scottish Studies Archives, University of Edinburgh (PN1978.02).
35. Presbytery of Mull Minutes, 1744. Argyll CC Archives. Re alleged incest between Mr. Frances MacDonald, Presbyterian Minister in Strontian and his sister.
36. *RMS Vol. VII, 272,* 1610. Register of the Great Seal of Scotland..
37. Maclean,-Bristol, Nicholas (Editor), 1998. *Inhabitants of the Iner Isles, Morvern and Ardnamurchan, 1716.* Scottish Records Society, New Series, Vol. 21. Edinburgh.
38. Grant, Francis J. Ed. (1909). *The Commissariot of Argyll: Register of Inventories, 1693-1702.* Scottish Records Society; James Skinner & Co, Edinburgh.
39. Macfarlane, Walter, *Geographical collections relating to Scotland, Vol. 2.* Scottish History Society, 1906-08.
40. 1695 Retours (Argyll) No.93.
41. Admiralty Chart 2507. *Ardnamurchan Point to Loch Bhreatal, Skye.* Surveyed 1852/3 by Capt. H.C. Otter and Commander J. Wood, R.N.
42. John Thomson (1832). *Atlas of Scotland. Map of North Argyll.* NLS.
43. William Roy's *Military Survey of Scotland, 1748-54.* British Museum.
44. NRS: GD345/571 f.83.
45. NRS: CC2/7/50/1. Summons by the Duke of Argyll of tenants in Ardnamurchan for debt, 1734.
46. Morison, William (1778 (?). *Report of the Contents and Estimate Rents of part of the Annexed Estate of Lochiel.* Taken June 1778 (?). Copy by Dr. Chris. Robinson in Fort William Library.
47. Hugh MacKenzie (1978). School of Scottish Studies Archives, University of Edinburgh. (PN 1978. 03).

48. Tobermory Court Records, 1870; Ref. 2. Argyll County Council Archives, Lochgilphead.
49. Mary Cameron; Oral Tradition, 2011.
50. Nicolaisen, W.F.H ((1982) 'Scandinavians and Celts in Caithness: The Place-Name Evidence', in J.R. Baldwin (Ed). *Caithness: A Cultural Crossroads,* 1982; 75-85.
51. Ordnance Survey Ist Edition Maps.
    Argyllshire, Sheet XV, Surveyed 1872, published 1875.
    Argyllshire, Sheets XIV, XVI, XXIV, XXV; Surveyed 1872 by Capt. Bolland RE, published 1875.
52. Campbell, Herbert, (1933). *Argyll Transcripts. Abstracts of the Particular Register of Sasines for Argyll* Vol. II (1st Series); No. 344, 25 Nov. 1651. Edinburgh.
53. NRS. SC59/2/11. Sheriff Court Processes (Tobermory), 1852.
54. Duncan Cameron, School of Scottish Studies Archives, University of Edinburgh. PN1966.118.
55. Rentalia Domini Regis, Appendix, pp. 622-625. (1541).
56. Steer, K.A. and Bannerman, J.W.M. (1977). *Late Mediaeval Monumental Sculpture in the West Highlands.* RCAHMS, Edinburgh.
57. Black, R (2009). 'Ballimore's Dream' in *West Highland Notes & Queries, Series 3, No. 14, 19-27, December 2009.* The Society of West Highland and Island Historical Research.

Skerries off Rubha Saune

INDEX OF WEST ARDNAMURCHAN COMMUNITY COUNCIL AREA
PLACE-NAMES AS RECEIVED
TOWNSHIP NAMES IN BOLD

INDEX OF GAELIC NAMES IN THE WEST ARDNAMURCHAN COMMUNITY COUNCIL AREA
ASTERISK DENOTES HYPOTHETICAL FORMS
TOWNSHIP NAMES IN BOLD. NORSE NAMES MARKED (ON)

Bronze Age ring cairn on Ardslignish, overlooking Loch Sunart

THE LOST PLACE-NAMES OF

MOIDART

SET IN A HISTORICAL, ARCHAEOLOGICAL

AND CULTURAL CONTEXT

CONTENTS

Drop from a mediaeval crozier found on the old coffin route between Dalelia and Eilean Fhìonain   © National Museums Scotland

Title Page: Rubha a' Chladha on Annat, with the burial ground on the headland, the fank, a dyke following the Annat Burn down to the burial ground enclosure, and creel buildings on both sides of the burn.

MOIDART TOWNSHIPS in 1739

Lands in Moidart belonging to MacDonald of Clanranald; life-rented to the Dowager Lady Clanranald, 1739.

Lands in Moidart pertaining to Donald MacDonald of Kinlochmoidart

Lands in Moidart belonging to MacDonald of Glenalladale

Lands in Moidart wadset to MacDonald of Dalelia

Lands in Arisaig belonging to MacDonald of Belfinlay

MOIDART TOWNSHIPS

A PROVISIONAL PLAN

Arisaig

Glen Finnan

Sunart

Ardgour

Arisaig

Loch nan Uamh

Ardnish

Loch Eilt

Muic

Creigvrodinn

Salnoes

Glen-Aladale

Essan

LAND OF ARISAIG

Brynafern

Allt an Dobhrain

Dalroyd

Loch Shiel

Arean

THE 24 MERK

Ulgary

Doilet

Annat

Kina-Creagan

Easter Alisary

Assary

PART OF

Inchrory

Glen Forslan

Drumloy

Wester Alisary

Leid na Cloich

Lochans

Dalelia

Loch Ailort

Brunary

Nether Kinloch

Langal

Aorinn

Upper Kinloch

Bad na Croggan

Mingary

Loch Moidart

Port a' Bhata

Blain

Glenuig

Kylesmore and Kylesbeg

Scardoish

Briaig

Samal-aman

Egnaig

Shona Beg

Smirisary

Baille

Arean

Tongaol

Sunart

Ardnamurchan

N

Scale in Km

0 1 2 3 4 5 6

166

LANDS IN MOIDART BELONGING TO MACDONALD OF CLANRANALD: LIFE-RENTED TO THE DOWAGER LADY CLANRANALD, 1739.

TONGAOI : Tòn na Gaoithe : the lower township

1718, Tonguy (6); 1739, Tongaoi (1); 1748, Tongorvie (32); 1760, Tongey, Island of Shuna (43); 1788, Tonguie (39); 1813, Island of Shona or Shonavore (49); 1823, Island of Shona (7); 1832, Ton Guich (34); 1844, Isle of Seona (29); 2002, Bailetonach (2).

Valuation: A 2 farthing land in 1748, a ½ merk land in 1718 and an 8 chilicks land in 1760 (43).

Tenants: 1718, Ranald McDonald (an old man) and Angus McRory, with half the township lying waste (6); 1746, Donald McDonald, John MacLeod, Angus MacVorrich, Calum MacGawry (12); 1748, Angus McIsaac and his father John (who was old and infirm) each had ⅝ of a farthing land , and Angus McInnes who rented ¾ of a farthing land (32); 1760, Angus McIsac (3 chilicks), Mary McDonald (3 chilicks) and John McIsac (2 chilicks) (43); 1788, Tack between John Macdonald in Ardnafuran and Andrew Macdonald; 1813, Eilean Shona was sold to Alexander Macdonald of Glenalladale (49).

Settlements: NM 6365 7355. *Bailetonach.* A scattered group of at least 25 structures.
                       NM 6294 7355. *Aonach.* A group of 6 structures.
                       NM 6420 7487. *Baramore.* Three structures and a small field system.

Shielings: NM 6385 7385.
                NM 6391 7364.

Placenames:

1. **Tongaoi (1). NM 6390 7340.** G. *Tòn na Gaoithe* or *An Tòn Gaoithe* '(the) windy bottom' or literally, 'the bottom of the wind', with *tòn* m. and gen. of *gaoth* f. 'wind'. The prevailing wind sweeps up through the pass here.

2. **Baille Tonach (2). NM 6390 7340.** G. *Am Baile Tònach* 'the bottom village', with *baile* m. and an adjectival form of *tòn,* possibly based on the name *(An) Tòn Gaoithe.* (see 1. above).

3. **Ru Ard Nie (3). NM 6338 7557.** G. Possibly *Rubha Aird Nì* 'the point of the stags (or herd of deer) height'. It is unlikely to be *Rubha Àird an Nighe* or *Nighidh,* 'the point of the headland of the washing or bathing', with *rubha* m., *àird* f. and gen. of *nighe(adh)* f. 'washing, cleaning, bathing'. This is a low, rocky headland backed by a precipitous hill, with difficult access from any of the settlements except by boat, and it is not the most accessible or the safest place to wash or bathe. It is also unlikely to be G. *nì,* 'cattle', as there is no grazing land in the vicinity. *Rubh Aird an Fheidh on O.S. maps.*

Sgurr an Teintein and *Rubha Àird an Nighe* from the south end of Smirisary

4. **Sgur na Hinagh (3); Sgur na Rinagh (34). NM 6365 7496.** G. Possibly *Sgùrr na h-Ìghneig* 'the peak of the young girl', with gen. sg. of *nighneag* f., diminutive form of *nighean* f. 'girl'. Named *Sgùrr an Teintein* 'the peak of the fire'

on OS maps, or more correctly *Sgòr nan Teintean* 'the peak of the fires', alias *Sgòr an Teine* 'the peak of the fire', with gen. pl. and sg. respectively, of *teine* m. 'fire'. This is a superb vantage point and a good location for a beacon fire. 'Shona' is derived from ON *Sjón ey* 'the island of watching', so perhaps this is the lookout point from which the island name was ultimately derived, and the site of an associated warning beacon.

5. **Id na Corrie More (3). NM 6254 7395** or **NM 6260 7400.** G. *Eilean a' Choire Mhòir* 'the island of the big whirlpool', with gen. sg. of *coire* m. 'whirlpool'. The name probably applies to the largest island in the group, which is un-named on any of the maps; the other islands are little more than skerries.

6. **Cruach Corrie More (3). NM 6315 7395.** G. *Cruach a' Choire Mhòir* 'the crag of the big whirlpool. Named *Cruach a' Choire* on OS maps.

7. **Bou Carrage (3); I. Bou Carrach (34)). NM 6243 7388.** G. Probably *Bodha a' Coire* 'the sunk rock of the whirlpool', with *bodha* 'a rock over which the waves break'. Possibly *(Eilean) Bodha Carragh* or *Bodha a' Charragh* 'the submerged reef of the rock', with gen. sg. of *carragh* m., 'rock'. Named *Eilean a' Choire* 'the island of the whirlpool' on OS maps.

8. **Id m Onich (3); Id. M Unich (34). NM 6285 7234.** G. Possibly *Eilean an Omhanaich* 'the island of the foamy place', with gen. sg. of *omhanach,* based on *omhan* m. 'froth'. The waves beating on the rocky shore probably produce a lot of foam here. Possibly *Eilean an Aonaich* 'the island of the hill pasture or moor'. The adjoining settlement is Aonach, so 'Aonach Island' is a possibility.

9. *I. Rhuall (3).* **NM 6280 7286.** G. *Eilean Raghnaill* '*Raghnall*'s Island', with gen. of the man's name *Raghnall* (formally spelt *Raonall*) 'Ronald'. Named *Eilean Raonuill* on OS maps. In 1653, Ranald MacDonald, thirteenth of Clanranald married his cousin Anne (or Agnes), daughter of John MacDonald, tenth of Clanranald. *'There is a dispensation 'dated at **Ellan Raald**, the 8ᵗʰ June, 1653,' granted for this marriage by "Dominicus Dingin," under authority of a commission from the Pope, "to dispensate in such business," written on the 10ᵗʰ of December. 1651. The parties are described as "in the second and third degree of consanguinity," whereas all mariages "contracted within the fourth degree, inclusively, are, by the universal Church of God, prohibited and declared of no force or value without a special dispensation from the said Church".* MacKenzie (1881).

10. **Bou an Taylor (3). NM 6262 7455.** G. *Bodha an Tàilleir* 'the tidal rock of the tailor', with *bogha* m. 'submerged /tidal rock or reef'; and the gen. sg. of *tàillear* m. – although the specific might have been a man's nickname, indeed the name is conceivably *Eilean 'an Tàilleir* 'the island of Iain the Tailor', with gen. of a man's name, *Iain*, 'Ian, John', shortened to '*an Bodha* is derived from ON *boði* 'breaker; sunk rock in the sea'.

11. **Sgurr an Tachan (4). NM 6393 7465.** G. *Sgùrr an Taghain* 'the peak of the pine marten', with gen. sg. of *taghan* m. 'pine marten'.

12. **The Park (4). NM 6378 7339.** This was probably the main field of the township.

13. *Allt Creach (4).* **NM 6360 7378.** Possibly G. *Allt na Crìche* or *Allt nan Crìoch* 'the burn of the boundary or boundaries, or the march burn(s)', with gen. sg. or gen. pl. of *crìoch* f. 'boundary'; alternatively, this may simply be *Allt na Creach* 'the burn of the crags', with gen. pl. of *creach,* perhaps for *creag* 'rock, crag' or cf. *creachann* 'crag, mountain'. Named *Allt an Tàilleir* 'the burn of the tailor' on OS maps, with gen. sg. of *tàillear* m. 'tailor'.

14. *Cnoc Coire (4).* **NM 6318 7395.** G. *Cnoc a' Choire* or *Cnoc nan Coire* 'the hill of the corrie(s)'. with gen. sg. or gen. pl. of *coire* m. 'corrie', or perhaps more likely, 'the hill of the whirlpool', as it overlooks *Eilean a' Choire* and *An Coire*. Named *Cruach a' Choire* on OS maps.

15. *Bogha na Leach (4).* **NM 6303 7313.** G. *Bodha nan Leac* 'the tidal rock of the flagstones', with gen. sg. of *leac* f. 'flagstone, flat rock'.

16. *Sgeir Leac Mòr (4).* **NM 6316 7319.** G. *Sgeir nan Leac Mhòr* 'the greater S*geir nan Leac* (the skerry of the flagstones)'.

17. *Sgeir Leac Beag (4).* **NM 6300 7330.** G. *Sgeir nan Leac Bheag* 'the lesser *Sgeir nan Leac* (the skerry of the flagstones)'.

18. *Rudha Àrd a' Bhag (4).* **NM 6270 7352.** G. *Rubha Àird a' Bhag* 'the point of the headland of the bag', from G. *baga* 'bag', or less likely G. *bac* 'slope, bank'.

19. *Sgeir Coire (4).* **NM 6243 7386.** G. *Sgeir a' Choire* 'the skerry of the whirlpool', with gen. sg. of *coire* 'whirlpool'. The 1:25,000 OS map names this *Eilean a' Choire*, which is the name of the island 100m to the east.

20. **Bogha Leachach (4). NM 6246 7412.** G. *Am Bodha Leacach* 'the flat submerged reef', i.e. consisting of flat rocks, with the adj. *leacach,* based on *leac* f. 'flagstone, flat rock'.

21. **Howarth Rock (4). NM 6180 7416.** A skerry to the west of Eilean Shona which is likely to be hazardous to shipping entering both the North and South Channels. Possibly named after Lord Howard of Glossop, the estate owner in the 19th century.

22. **Bogha Mor Carn nan Taillier (4). NM 6263 7458.** G. *Bodha Mòr Càrn an Tàilleir* 'the large submerged reef of *Càrn an Tàilleir* (the cairn of the tailor), with gen. sg. of *tàillear* m. 'tailor'. The 1st Edition OS 1:10,560 map shows *Bogha Mòr an Tàilleir* 'the sunk rock of the tailor' at NM 6260 7400.

23. **Carn na Biast Due (572 ft) (4). NM 6366 7496.** G. *Càrn nam Biast-dubha* 'the cairn of the otters', with gen. pl. of *biast-dubh* f. 'otter'. Named *Sgurr an Teintein* (176m), on the OS 1:25,000 map.

24. **Du Skeir (4). NM 6318 7567.** G. *An Sgeir Dhubh* 'the black skerry', with *sgeir* f. 'skerry'.

25. **Rudha Ard na Fiann (4). NM 6378 7527.** G. *Rubha Àird an Fhèidh,* alias *Rubha Àird nam Fiadh* 'the point of the headland of the deer', with gen. sg. or gen. pl. of *fiadh,* m. 'deer'. The documentary form might be for G. *Rubha Àird nam Fiann,* literally 'the point of the headland of the warriors', with gen. pl. of *fiann* m. 'warrior, giant'. The OS 1:25,000 map shows *Rubh' Àird an Fheidh* and the 1:10,560 1st Edition *Rudh' Àird nam Fiadh* 'the height of the deer promontory' at NM 6340 7577.

Birlinn Chlann Raghnaill on Clanranald's armorial bearings dated 1641 in *Cill Ma-Ruibha* 'Kilmorie', Arisaig.

26. **Port Uidheamachadh Fada (17). NM637 733.** G. *Port Uidheamachaidh Fada* 'the long (i.e. deeply indented) *Port Uidheamachaidh'* or *Port an Uidheamachaidh Fhada,* literally 'the landing place of the long preparation', i.e. 'a landing place where boats were equipped', both with gen. sg. *uidheamachaidh* m. 'equipping, preparation'. This could be where Clanranald's birlinns were fitted out. The illustration above, from the Clanranald armorial bearings of 1641 in *Cill Ma-Ruibhe* 'Kilmorie', Arisaig, depicts a change in birlinn design that seems to have taken place after about 1550.

27. **Ton Goath (4). NM 6343 7358.** G. *Tòn Gaoithe* or *Tòn na Gaoith* '(the) windy bottom', literally 'the bottom of the wind', with gen. of *gaoth* 'wind'.

28. **Slandrhul (36). NM 6280 7286.** G. *Eilean Raghnaill* 'Raghnall's Island', gen. of the man's name *Raghnall* (formerly written *Raonall*) 'Ronald'. Named *Eilean Raonuill* on OS maps and described by William Bald as belonging to Clanranald. It is big enough to graze one or two sheep or goats on, but there is probably more seaweed than grass. Like No. 9 above, it is a phonetic rendering of *Eilean Raghnaill.*

29. **Ba Vare (34). NM 6430 7490.** G. *Bagh Bharra* 'Barr bay'.

30. **Ceann Shios (19). NM 6294 7355.** G. *Ceann Shios* 'the westmost' or 'the west end'. A house at the western end of the township, and last occcupied by the Kennedy family. It appears to have been a colloquial name for Aonach, and occurs in the Mingary Baptismal Register.

AREAN : Àirigh Iain : Iain's Shieling

1697, Ariran in Mudart (58); 1718, Arian, Arrean (6); 1718, Ariean (61); 1739, Aryian (1); 1746, Aryean (12); 1748, Ari-Ian (32); 1760, Ari Ian, (43); 1782, Arian (64); 1788, Ongaw (39), (39); 2002, Arean (2).

Valuation: A ½ merk land in 1718 and 1760, and a 2 farthing land in 1748.

Tenants: 1697, Angus McCallum, his spouse Sisilie McNeale and their children, Donald mor and Donald Oig (58); 1718, Dougal McDonald alias McEan vic Coull, Donald More McInish and Marion McDonald (Widow) with part of the township lying waste (6); 1744, Donald McDonald and Christy McVarish his wife (38); 1748, Dugald MacDonald, ¼d land and Rory MacPherson in 1748 (12); 1749, Dugald McDonald and Donald McDonald each possessed a farthing land (32); 1760, Donald McDonald Senior and Donald McDonald Junior equally between them (43); 1782, Donald MacMartin alias Cameron (64);1788, Tack between John Macdonald in Ardnafuaran and Andrew Macdonald (39).

Settlements: NM 6453 7301. *Arien.* Four structures.
NM 6468 7497. *An Dubh-aonach.* Five structures.
NM 6482 7473. '*Allt an Dubh-aonach*'. Three structures.
NM 6396 7332. One structure recorded.
NM 6422 7441. One structure recorded.

Shielings: NM 6494 7300. *Airidh Iain Iosal.*

Placenames:

Iosal 'n Uachdar' – Lower and Upper

1. **Arean (2). NM 6453 7301.** G. *Àiridh Iain* 'Iain's shieling', with gen. of the man's name, *Iain* 'Ian, John'.

2. **Rhu Rhapple Bainn (3). NM 6486 7268.** G. *Rubha a' Chapaill Bhàin* 'the point of the white horse', with gen. sg. of *capall* m. 'horse'. Spelled **Ru Chaple Bann** on Ms 384 (1).

3. **Ru an Daunach (3). NM 64547514.** G. *Rubha an Dubh Aonaich* 'the point of *An Dubh Aonach*' 'the point of the black hill' or 'black (poor?) hill pasture'.

4. **Port an Dounach (3). NM 6454 7492.** G. *Port an Dubh Aonaich* 'the port of *An Dubh Aonach*'; see no. 7, below.

5. **Caolas a' Tuath (4). NM 6468 7515.** G. *An Caolas a' Tuath* 'the northern strait'. The 'Narrows' here are only 60m wide.

6. **Cnoc an Samhrach (4). NM 6449 7391.** ?G. *Cnocan Samhrach* 'the hill of the summer grazing'. *Cnocan*, gen. *cnocain* m. 'hillock'; *samhrach* adj. 'pertaining to summer'. Possibly a shieling site. If this was an ancient woodland site, then *Cnocan nan Sòbhraig* 'the hillock of the primroses' is a possibility, with *sòbhrach/sòthrag* 'primroses'.

7. **Duanachd (4). NM 6466 7496.** Probably G. *An Dubh Aonach* 'the black hill' or 'black (poor?) hill pasture'.

8. **Uillt a Bhara (4). NM 6410 7451.** G. *Uillt a' Bharra* 'the stream of the summit', with gen. sg. of *bàrr* m. 'summit'. Named Baramore Burn on OS maps (2), with *Barr Mòr* 'big headland'.

9. **Allt na Davan (4). NM 6460 7450.** G. *Allt an Da(m)bhain* 'the burn of the spider', with gen. sg. of *damhan-(allaidh)* m. 'spider', or 'the burn of the ox', with gen. sg. of *damhan*, as a diminutive of *damh* m. 'ox'.

10. **Clach na Creachan (4). NM 6430 7463.** ? G. *Clach na Creachainn* 'the stone of the mountain', with gen. sg. of *creachann* f. 'mountain, rocky mountain top'. *crest*

11. **Right Daroch (4). NM 6410 7423.** G. *An Ruigh Daraich* 'the oak shieling- ground', with gen. of *darach* m. 'oak'. Shown as a hill on the chart but the name implies an area of summer pasture.

12. **Druim Beith (4). NM 6414 7390.** G. *An Druim Beithe* 'the birch ridge', with gen. sg. of *beith* m. 'birch'.

13. **Rudha na Clagha Creag (4). NM 6465 7514.** G. *Rubha na Cloiche Crèadha* 'the promontory of the clay rock', with gen. sg. of *clach* f. 'rock'. Named *Rubh' an Dubh Aonaich* 'the headland of the black ridge' on OS maps (2).

14. **Port na Creaghan (4). NM 6475 7490.** ? G. *Port nan Creagan* 'the landing place of the rocks', with gen. sg. of *creagan* 'rocks'.

15. **Cnoc na Ath (4). NM 6442 7319.** G. *Cnoc an Àtha* 'the hill of the ford', with gen. sg. of *àth* m. 'ford'.

Cnoc na L'Atha – Hill of the Kiln – any evidence?

16. **Port Ard na Fhraoich (4); Ru A Freugh (3). NM 6425 7292.** G. *Port Àird an Fhraoich* 'the landing place of the headland of the heather', with gen. of *fraoch* m. 'heather'. The OS has *Rubha Àird an Fhraoich* 'the headland of the heathery height' at NM 6410 7296, but the headland is hardly high enough to be a 'height'.

17. **Airidh Iain Iosail (4). NM 6494 7300. G.** *Àiridh Iain Ìosal* 'the lower *Àirigh Iain* (*Iain*'s shieling)'.

18. **Port an Sluichd (4). NM 6497 7277.** G. *Port an t-Sluic* 'the landing place of the hollow', with gen. sg. of *sloc* m. 'hollow'.

19. **Creag na hullar (34). NM 6469 7465 (?).** G. *Creag na h-U...? possibly G.iolair* 'eagle' or iobhar 'Yew Tree'

Eilean Shona from Kylesmore

BAILIE : Am Baillaidh : the township or chief residence
1718, Ballich (6); 1718, Baillich (61); 1748, Bailie (32); 1760, Bailey (43); 1788, Ballie (39); 1816, Bailie (3).

Bailie from Eilean Tioram

Valuation: A ½ merk land in 1718, a 2 farthing land in 1748 and 8 chilicks in 1760. A chilick or clitick from Gaelic *cleitig*, *clitinn* or *clitig* equals a ½ farthing land. Anything smaller would be a *sianag*.

Tenants: 1718, John McRory (¼m) and Rory McEan vic Coul (¼m) (6); 1744, John McIsag or McQueen and Marion McDonald his wife resident on Eilean Shona, and Mary McDonald from Ardnamurchan (38); 1748, Donald Mor MacDonald and his brother Iain Og vic Ruari (12); 1748-1790, Seumais mhic Dhomhnuill Mhoir mhic Alasdair (son of Donald Mor), aka Seumas a' Bhaillaidh mac Dhomhnuill Mhòr mhic Alasdair mhic Ruaridh III of Glenaladale and his mother Mary each held a farthing land (12, 32); 1760, James McDonald (5 chilicks) and Angus McIsac (3 chilicks) (43); 1788, Tack between John Macdonald in Ardnafuaran and Andrew Macdonald (39).

Settlements: NM 6551 7298. *Am Baillaidh*. Now Eilean Shona House and 'village'. Three structures on the estate plan.
NM 6565 7363. No name. Three structures.
NM 6583 7382. 'Doireaneighinn'. Three structures.

Shielings: NM 6490 7456, *Àiridh Iain Aird*.

Placenames:
1. **Baley (1). NM 6548 7300.** G. Possibly *Am Baillaidh* 'the chief residence or township'. Possibly elliptical for G. *Taigh a' Bhàillaidh* 'the house of the factor', or similar, with gen. sg. of *bàillidh* m., 'factor, baillie', but this is unlikely. cf. nos 7 and 25, below.

2. **Slochk (3). NM 6497 7280.** G. An *Slochd* 'the hollow'.

3. **P. more (3). NM 6560 7298.** G. *Am Port Mòr* 'the large landing place'.

4. **P. an Grey (3). NM 6580 7323.** G. *Port na Creidhe* 'the port of clay', with gen. sg. of *criadh* f. 'clay', assuming this is suitable topographically; see also no. 17, below. Named **Purt an Cray** on Ms 384 (1).

5. **Port an Tarbert (3). NM 6615 7360.** G. *Port an Tairbeirt* 'the landing place of the isthmus' or 'port of the portage place', with gen. sg. of *tairbeart* f., 'isthmus', but often implying 'a portage'.

6. **Derry Nian (4). NM 6591 7403.** G. *Doire an Eidhinn* 'the groves of the ivy', with gen. of *eidheann* m. (elsewhere f.) 'ivy'. Named Doireaneighinn on OS maps

7. **Ben e Valey (3). NM 6532 7343.** G. *Beinn a' Bhaillaidh* 'the ben of the residence', with gen. sg. of *baillaidh* m. 'residence'; cf. nos 1, above and 25, below. The name applies to the shoulder of the hill overlooking the settlement.

8. **Rudh a Bothan (4). NM 6590 7406.** G. *Rubha a' Bhothain* or *Rubha nam Bothan* 'the point of the bothy or bothies', with gen. sg. and gen. pl. of *bothan* m. 'bothy'.

9. **Leacann (4). NM 6542 7360.** G. *An Leacann* 'steep shelving ground' or 'slabby rocks'. Shown as a small settlement on the chart.

10. **Cnoc an Sluichd (4). NM 6509 7300.** G. *Cnoc an t-Sluichd* 'the hill of the hollow', with gen. sg. of *slochd* m. 'hollow'.

11. **Cnoc an Sitheacan (4). NM 6500 7329.** G. Possibly *Cnoc nan Sìtheachan* 'the hill of the fairies', with gen. pl. of *sitheach* m. 'fairy', if not *Cnoc nan Sìthichean* with the same meaning, with gen. pl. of *sìthiche* m. 'fairy'.

12. **Allt a Dail (4). NM 6544 7300.** G. *Allt na Dail* 'the burn of the field' with gen. sg. or pl. of *dail* f. 'field, meadow'.

13. **Tigh Bhaile (4, 19). NM 6552 7293.** G. Possibly *Taigh a' Bhaillaidh* 'the house of the residence', with gen. sg. of *baile* m. 'village, township or residence', or 'Baillaidh House'. cf. nos 1 and 7, above, and 27, below. On the site of Eilean Shona House. Seumas a'Bhaillaidh (James MacDonald) , 1730-1796 was tacksman here before emigrating to Canada in 1790 (see 'tenants, above).

14. **Cnoc an Feur (4). NM 6542 7324.** G. *Cnoc an Fheòir* 'the hill of the hay or grass', with gen. sg. of *feur* m. 'grass'. Here perhaps, a 'grassy hillock'.

15. **Allt na Srath (4). NM 6562 7325.** G. *Allt an t-Sratha* or *Allt nan Srath* 'the burn of the valley or valleys', with gen. sg. of *srath* m. 'valley, strath'. Named *Allt na Crioche* on the 1st Edition OS map. G, *Allt na Crìche* 'the burn of the march / boundary', with gen. sg. of *crìoch* f. 'boundary'.

16. **Porst na Lamrig (4). NM 6560 7298.** G. *Port na Lamraige* 'the landing place of the channel', with gen. sg. of *lamraig* f. 'landing place or clear channel'.

17. **Porst na Greath (4). NM 6580 7323.** G. *Port na Creidhe* 'the landing place of the clay'; see no. 4, above.

18. **Torr am Bainne Iam (4). NM 6586 7364.** Perhaps in error for 'Torr am Bainne Lom' i.e. G. *Tòrr a' Bhainne Luim* 'the hill of the skimmed milk', with gen. sg. of *bainne lom* m. 'skimmed milk'.

19. **Allt Ghoire an Fiann Pheanner (4). NM 6552 7408.** G. *Allt Coire ?*. The meaning of the name is obscure. The Chart shows this as a prominent burn but none of the OS maps suggest a burn anywhere near here.

20. **Airidh Iain Ard (4). Àiridh Iain Uachdarach (19). NM 6490 7456.** G. *Àiridh Iain Àrd, Àiridh Iain Uachdarach* 'the upper *Àiridh Iain* (Iain's shieling)'; cf. *Àiridh Iain* and *Àiridh Iain Ìosal* in Arean.

21. **Creag na nullar (3). NM 6490 7415.** G. Possibly *Creag na h-Iolaire* 'the creag of the eagle', with gen. sg. of *iolaire* f. 'eagle'. Possibly *Creag nan Ealar* 'the rocks of the salt (pans ?)'.

22. **Sgudallan (4). NM 6575 7414.** Possibly *An Sgudalan* 'the rubbish or offal tip', with *sgudal* m. 'rubbish, offal', plus suffix of place – *an*. Possibly just flotsom on the tide-mark.

23. **Porst Gor a Fan (4). NM 6588 7395.** G.? *Port a' Ghobhair Bhain* 'the port of the white goat', possibly a rock that looks like a white goat. The form 'Gor a Fan' might be for G. *garbhan* m. 'fish gills; coarse (oat)meal'; perhaps G. *Port nan Garbhan* 'the landing place of the gills' or *Port a' Gharbhain* 'the landing place of the meal', with gen. pl. or gen. sg. of *garbhan*. Alternatively, *Garbhan* may have been a name in its own right, based on *garbh* 'rough' plus suffix of place –*an; Port Gharbhain* 'the landing place of *Garbhan', 'the rough landing place', with gen. sg. of *garbhan*.

24. **North Harbour (4). NM 6530 7470.**

25. **Ru Vey (13). NM 6555 7273.** G. *Rubha a' Bhàillidh* 'the point of the residence' as on OS maps, with gen. sg. of *bàillidh* m. 'residence'; cf. no 7, above. Possibly *Rhu' Bheithe* 'the point of the birches'.

26. **Beinn a Beile, 863 ft. (4). NM 6490 7416.** Named *Beinn a' Bhàillidh 265m* on OS maps, which may be correct as *beile* 'bridle-bit' seems unlikely.

SCARDOISH : Sgàird an Deagh Uisge : the pass of the pure water

1718, Scardoish (6); 1744, Scardoosh (38); 1744, Scardosh (38); 1748, Scardeoisk (32); 1760, Scardoish (45); 1782, Scardosge (64); 1798, Scardosie (71); 1816, Scardossie (3); 1823, Scardossie (7).

Cruach nam Meann and *Cùil Doirlinn* from *Eilean Tioram*

Valuation: A ⅝ merk land in 1718, a 3 farthing land in 1748 and a 2½ farthing land in 1760 and in 1798.

Area: C. 1816 there was 31.68 acres of arable in 6 fields, Eilean Tioram with some arable land, totalling 4.60 acres and 166.64 of moor, a total of 202.92 acres. The Island of Risga appears to have been let with Scardoish, and measured 17.62 acres, but it was planted with trees c. 1813.

Tenants: 1718, John McGillespick (Ground Officer) ¼m, David Paterson (Mason) ¼m and Patrick McGilandrish, ⅛ m. William Paterson, John Macleasback, Patrick MacGillendrich (6); 1744, John McNeill, John McMillan and Mary McInnes living in Tyrim (38); 1744, Angus McQueen and his wife Mary McVarish, Donald McPherson (38); 1745, John Mac Leasback, Patrick MacGillendrich and William Patterson (1748, Patrick McPherson, ½ farthing, Angus McIsaac, ½ farthing, Katherine McVarish, ½ farthing, Katherine McVarish, ½ farthing, John McInnes, ¼ farthing, Mary McDonald, ¼ farthing and ½ farthing lying waste (32); 1760, Ronald mc Pherson, Angus mc kisaig, John Patterson, Archibald mc kisaig and Angus mc Donald held the 2½ farthing land equally between them (45); 1782, Duncan MacIsaak (64); 1798, Donald Macisaac and 4 others (71).

Settlements: NM 6648 7140. *Scardoish.* 8-10 buildings on the estate plan.

Shielings: Not known, but it was probably too small a township to warrant having a shieling, other than perhaps Risga.

Placenames:

1. **Scardossie (3). NM 6648 7140.** G. *Sgàird an Deagh Uisge* 'the pass of the pure water'. Possibly G. *Sgairt Dheabh-Uisge* 'the pass of the water that dries up', with *Sgairt* derived from ON *skarfr* 'a pass' or 'an opening cut among the rocks', which it most certainly is. Pronounced *Sgairt 'eabh-uisge* locally.

2. **Stron Laung (3). NM 6663 7262.** G. *Sròn nan Long* 'the point of the ships' with gen. pl. of *long* f. 'ship'.

3. **Ru Crochry (3). NM 6625 7195.** G. *Rubha a' Chrochaidh* 'the point of the hanging', with gen. sg. of *crochadh*, m. 'hanging'; cf. OS 1875 *Cnoc a' Chrochaidh*.

4. **Drian Dubh (3). NM 6626 7069.** G. *Droigheann Dubh* 'the blackthorn thicket'.

5. **Cruoch Bennan (3). NM 6690 7160.** The current map form is G. *Cruach nam Meann* 'the stack of the kids', with gen.pl. of *meann* m. 'kid, goat'. The 'lost' form may be for *Cruach Bheinn* or *Cruach Bheannan* 'stack mountain'; *beannan* m. is a diminutive form of *beinn* f. 'mountain', or with *Bhinnein* 'pinacle'. Either name may possibly have derived from the other, by reason of popular etymology or transcription error. Alternatively, OS 1876 Beinn a Mhinn, i.e. G. *Beinn a' Mhinn* 'the mountain of the kid', with gen. sg. of *meann* m. may suggest that Cruach Benan may be a

174

typographical errror for *Cruach Beinn nam Meann* or *Cruach Beinn a' Mhinn* 'the stack of *Beinn nam Meann* or *Beinn a' Mhinn*..

6. **Sadil Scardossie (3). NM 6632 7144.** G. *Sàideal Sgàirt an Deagh Uisge* 'the bay of *Sgàirt an Deagh Uuisge'*. *Sàideal* does not appear in 'Dwelly', but there are several in Moidart, and each one is associated with a bay.

7. **Sron a' Chrochry (3). NM 6626 7116.** G. *Sròn a' Chrochaidh* 'the point of the hanging'. This is where Clanranald had his gallows. See no. 3, above.

One of the two Ice Houses at the Fishing Station

8. **Salmon House (4). NM 6615 7126.** Centred on the estate ice house close to where salmon were netted at the mouth of the River Shiel. Referred to in the Mingary Church Baptismal Register as **The Station.**

9. **Scor a Sgriach (4). NM 6635 7176.** G. *Sgor an Sgreucha* 'the cleft of the scream', with gen. sg. of *sgreuch* m. 'scream'. Perhaps by association with a echo with *Cnoc a' Chrochaidh* 'the hill of the hanging', 'Gallow's Hill'. See no. 3, above.

Casteal Tioram from the boat landing on the west side of the island

10. **Rudha Caistel Tioram (4). NM 6615 7242.** G. *Rubha a' Chaisteil Thioraim* 'the point of *An Caisteal Tioram*' (the dry castle).

11. **Cul na Doirlinn (4). NM 6642 7238.** G. *Cùl na Dòirlinn* 'the back of *Doirlinn*' i.e. *the spit*', with gen. sg. of *dòirlinn* f. 'isthmus'. The OS map (2) shows *Cul Doirlinn* on the south side of the island whereas the name would apply equally well to both sides of the sandspit. The Judicial Rental of the Estate of Clanranald dated 1798 states that *the Farm is Sandy-being so enwashed upon by the sea that it is now barely 30 acres excluding the rocks* (71). This refers to the field at the end of the causeway, which has been subjected to erosion by the sea, and past encroachment by sand, and it is still under threat today.

12. **Sgeir nan Sron Long (4). NM 6662 7265.** G. *Sgeir Sròn nan Long* 'the skerry of *Sròn nan Long*'. See no. 13, below.

13. **Sron na Long (4). NM 6680 7244.** G. *Sròn nan Long* 'the point of the ships', with gen. pl. of *long* f. 'ship'. Local Gaelic tends to drop the second '*n*' of '*nan*', hence 'Sron <u>na</u> Long'.

Riska from Sròn nan Long

14. **Beinnein an Eoin (4). NM 6672 7193.** G. *Binnein an Eòin* 'the pinnacle of the bird', with gen. sg. of *eun* m. 'bird'. *Binnein* (also *binnean*) m. is a derivative of *beinn* f. 'mountain'.

15. **Bentta corach (34). NM 6680 7102.** Possibly named *Beinn Gheur* by the OS.

16. **Changehouse of Castletirrim (7); Changehouse of Castle Tyrium (71). NM 6634 7235.** The inn on *Eilean Tioram* is at the end of the causeway, and shown but not named on the estate plan of c. 1816. See no. 17, below. Let to John Maceachan in Irine, 1798.

17. **The Drummer's Grave (23). NM 6634 7235.** According to Wendy Wood, 'this was an oblong of stones suggesting a cottage. The Drummer, probably one of Cromwell's men during his occupancy of the castle, was buried alive here'. This 'grave' could only be the Changehouse (16, above), but the building is very robust and probably had its origins in the late Mediaeval period.

18. *Àite suidhe Mhic ic Ailean (23).* G. *Àite-suidhe Mhic 'ic Ailein* '*Mac 'ic Ailein*'s seat' i.e. Clanranald's Seat'. Before Ailan MacDonald, 12th of Clanranald set out to join the Jacobite Army in 1715, he set fire to *Caisteal Tioram* to prevent it from falling into the hands of the Government forces. This is where he stopped and looked back to see the castle in flames.

19. **Fuaire Glac (75). NM 6635 7120.** G. *Fuaire Glac* 'the coldest hollow'. Putting the specific before the generic is unusual, and perhaps an old form, but the oral tradition insists that this is NOT *Glac Fuaire*.

Shiel Foot and the South Channel from Bealach Sgàirt an Dea-uisge

The mouth of the River Shiel from the jetty at the Fishing Station

BRIAIG : Brèidhig : the broad bay *? Breagh Eige? Beautiful notch?*

1700, Brinnaig (58); 1748, Breaig (32); 1760, Breaig (43); 1782, Briach (64); 1798, Briag (71); 1813, Breig (49); 1816, Brieg (3); 1823, Breig (7).

Briaig looking NW across Loch Moidart to Shona Beag and Eilean Shona.   © Dr. Sandra Evans

Valuation: A 1½ farthing land in 1748.

Area: C. 1816, there was 25.40 acres of arable in three fields (two very small and scattered) and 400.80 of moorland, a total of 426.20 acres.

Tenants: 1696, Christian N'Kissaig and her children Christian, Mary and Moir (58); Lying waste in 1748 (32); 1760, Archibald mac kisaag, late Ground officer in Moidart, the farthing land of Briaig (44); 1782, John and Donald MacIsaac (64); 1798, Duncan Macisaac and 3 others (71); 1813, sold to Alexander Macdonald of Glenalladale (49).

Settlements: NM 6768 7198. *Briaig*. At least 10 structures and a fank.

Shielings: NM 6956 7104. '*Lochan Glac Ealagain* '.

Placenames: *locally pronounced Breea Eek*

1. **Brieg (3). NM 6768 7198.** G. *Brèidhig* 'the broad bay'. Possibly a loan-word from the Norse *Breið-vik* 'broad- bay', and if so, the name has been transferred from the coast to the settlement.

2. **C Longurst (3). NM 6775 7232.** G. *Camas an Longphuirt* 'the bay of the harbour', with gen. sg. of *longphort* m. 'harbour'. The chart (4) shows the *Allt Camas Longart* discharging into the sea at NM 6800 7247.

3. **P an Dluigh (3). NM 6746 7233.** G. ? *Port nan Duilleag* 'the boat landing of the leaves'.

4. **Bealla Corrach (3). NM 6695 7118.** G. ? *Am Bealach 'Corrach'* 'the steep pass'. Perhaps with G. *corrach* 'steep, precipitous, abrupt, affording an insecure foothold'. Named *Bealach Sgàird an Deagh Uisge* 'the pass of *Sgàird an Deagh Uisge*' on OS maps (2). See Scardoish no. 1, 18.

5. **Coire (4). NM 6772 7217.** G. *An Coire* 'the corry'. A very small corry below the settlement.

6. **Innes Cruach Choire (4). NM 6763 7193.** G. *Innis Cruach a' Coire* 'the field of the hill of the corry'.

7. **Cruach a' Choire (4). NM 6716 7159.** G. *Cruach a' Choire* 'the hill of the corry'.

8. **Rudha na Baintichearna (4). NM 6735 7244.** G. *Rubha na Baintighearna* 'the headland of the lady'.

9. **Achi na Cuillidh (3). NM 6745 7068.** G. *Achadh na Cuillidh* 'the field of the horse' or 'of the small hollow'.

The lochan at the top of the Bealach Sgàirt an Deagh Uisge

Lochan na Fola

'Port an Dluigh', where birlinns could be drawn up for maintenance and overwintering

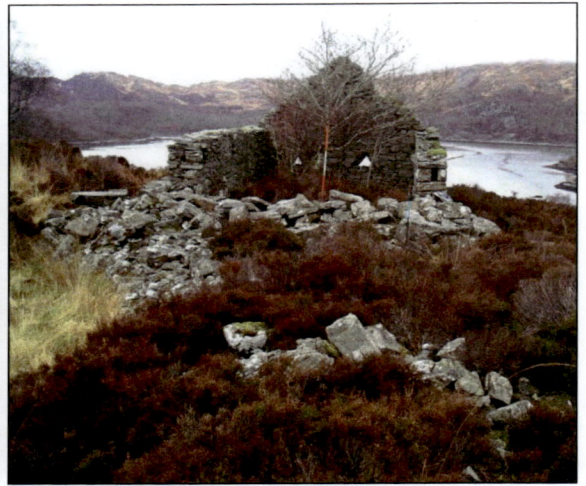

Houses on Briaig

A well-built barn from the SE

© Dr. Sandra Evans

Briaig corn kiln with veteran pollarded trees beyond

© Dr. Sandra Evans

BLAIN : Blàthan : green field

Flowers – Dwelly. Camus Bhlàthain + Blathaich

1718, Blayn (61); 1718, Blaym (6); 1739, Blauin (1); 1744, Blain (38); 1748, Blahain (32); 1760, Blain (44); 1796, Blan(71); 1813, Blain (49); 1816, Blain (3); 1823, Blain (7); 2002, Blain (37).

High Blain from the north-west and from the west

Valuation: a ½ merk land in 1718. a 2½ farthing land in 1748 and a 1½ farthing land in 1798.

Area: C. 1816 there was 10.52 acres of arable in three fields, 309.17 moss and 329.24 of moorland, a total of 648.93 acres.

Tenants: 1718, John McEan vic Ewn alias McEachen, with John, his father and Angus, his brother (6); 1744, John McEachin, Tenant (38); 1748, Alex. McEachan, ½ farthing, Alex. McDonald, ½ farthing, Donald McEchan, ¾ farthing and Angus McEchan, ¾ farthing (32); 1760, Archibald mac kisaag, late Ground officer in Moydart, a half farthing land, Donald Mac Eachin 3 cliticks and Alexander Mac Eachin 3 cliticks (44); 1798, John Maccachan and 2 others (71). Sold to Alexander Macdonald of Glenalladale in 1813 (49).

Settlements: NM 6750 6974. *Blain.* At least 10 structures shown on the estate plan.

Shielings: Not recorded.

Placenames:

1.   **Blain (1). NM 6750 6974.** G. *Blàthan* 'bloom, blossom', with diminutive of *blàth* m., perhaps in the sense 'green field', or G. *blian* 'a green spot'. Local tradition suggests that the name derives from a Gaelic word for 'devotion', with a possible association with St. Blane, an Irish monk whose monastic centre was at Kingarth, Bute.

2.   **Alt a Glenan (3). NM 6754 7000.** G. *Allt a' Ghleannain* 'the burn of the valley', with gen. sg. of *gleannan* m., diminutive of *gleann* m. 'valley'.

3.   **Ault Vach a Darroch (3). NM 6814 6968.** G. *Allt Bac an Daraich* '? the burn of the hollow of the oak'.

4.   **Ru Ard na Pall (3). NM 6805 6813.** G. *Rubha Àird nam Poll* 'the headland of the height of the pools', with gen. pl. of *poll* m. 'pool'.

5.   **Ru Ard na Darroch (3). NM 6857 6811.** G. *Rubha Àird an Daraich* 'the point of the headland of the oak', with gen. sg. of *darach* m. 'oak'.

6.   **Loch Achi na Culiah (3). NM 6734 7065.** G. *?Loch Achadh na Cuillidh* ' the loch of the field of the hollow or horse' with *achadh* gen. *achaidh* 'field'. Named 'Loch Blain' G. *Loch Bhlathain* on the OS Map (2). It is not clear from the Estate Plan if the loch and the field are on Briaig or on Blain, but the loch appears to be on Blain and the field on Briaig. The loch changed shape after the dam was built, and the field may well be at the head of the loch, c. NM 6728 7084.

7.   **Ru n Brachk (3). NM 6885 6834.** G. *Rubha nam Breac* 'the point of the trout', with gen. pl. of *breac* m. 'trout'.

181

The site of the house in the *Achadh na Cuillidh*

8. **Blainard** **(15). NM 675 698.** G. *Blàthan Ard* 'High Blain'.

Loch Blain from the east

MINGARY : Mioghairidh : from Old Norse Mikinn Garðr : Great Enclosure

1718, Meengary (6, 61); 1739, Myngarie (1); 1748, Mingary (28); 1760, Mingary (51); 1798, Mingarry (71); 1813, Mingary (49); 1823, Murgary (7); 2002, Mingarry (37).

Panoramic view looking towards Ben Resipole and Acharacle   © Dr. Sandra Evans

Valuation: A 1 merk land in 1718 4 farthing land in 1748 and a two farthing land in 1798

Area: C. 1816 there was 39.25 acres of arable land, 44.50 of moss and 304.87 of moorland, a total of 388.62 acres.

Tenants: 1718, John McCoilduy (3/16m), John McCoil vic Ewn (½m) and Donald McConihy vic Inish (¼m) with 1/16m lying waste (6); 1730, Donald McEan Vic Gowan alias McDonald, smith in Mingary (76); 1748, Archibald McIllick, ½ farthing, Hugh McVarish, ½ farthing, Mary McDonald, ¼ farthing, Mary McVarish, ¼ farthing, Roderick McPherson, ¾ farthing, Donald McIsaac, ¾ farthing and Donald Brown, 1 farthing (32). 1760, John Mac Donald (½ merk land), while Donald Bane Smith, Rhoderick Macpherson, Donald Smith and Donald MacIsack were each granted a ½ farthing land for 10 years, commencing June, 1760 (51); 1782, John Smith (64); 1798, Rory Macvarrish and four others (71). Sold to Alexander Macdonald of Glenalladale in 1813 (49).

Settlements: NM 6896 7041. *Mingary.*
NM 6890 6955. *'Cnoc a' Ghille Bhuidhe'*

Shielings: Not recorded.

Placenames:

1. **Mingarry (2). NM 6895 7042.** G. *Mioghairidh* from the Old Norse *Mikinn Garðr* 'the great enclosure'. Possibly G. *Mìn Ghearraidh* 'the smooth enclosure'.

Building on Mingary showing evidence of a change of use   © Dr. Sandra Evans

183

2. **Lochan Glac Dhealachan (?). NM 6960 7070.** G. Apparently *Lochan Glac Dhealachan* 'the loch of the hollow of the leeches', with gen. pl. of *deala* f. 'leech'. Possibly G. *Lochan Glac Dealachaidhean* 'the lochan of the hollow of parting'. The OS has G. *Lochan Glac Ealagain* 'the lochan of the hollow of the little swans', which seems more likely.

3. **Deeke (15). NM 6911 6909.** G. *An Dìg* 'the ditch' or 'the drain'. A croft to the east of Mingary, now known as *Mo Dhachaidh*.

4. **Cnoc a' Ghille Bhuidhe (15). NM 6899 6956.** G. *Cnoc a' Ghille Bhuidhe* 'the hill of the fair-headed boy'. A hundred metres to the east of this hill are the remains of some buildings and lynchets, but there is no local tradition of settlement here. The OS places this name at NM 7040 7059, on Langal.

5. **Loch an Glack Gellan (3). NM 6963 7020.** *Loch an Glac G...?* Possibly with *gealan* gen. *gealain* 'linnet', to give 'the loch of the linnet defile'.

Members of the *Comann Eachdraidh Mùideart* (The Moidart Local History Group) surveying Mingary as part of the 'Scotland's Rural Past' project.   © Dr. Sandra Evans

A bucht (cattle-milking pound) in a field near *Cnoc a' Ghille Bhuidhe*

PORT A' BHATA : Port a' Bhàta : the landing place of the boat
1798, Portavat (71); 1813, Portvait (49); 1816, Portuvat (3); 1832, Portuva (34); 2002, Port a' Bhata (37).

A 19th C. 'improved' house on Port a' Bhàta

Port a' Bhata appears to have formed part of Mingary at one time but by 1798 it was a township in its own right.

Valuation: A two farthing land in 1798 (71).

Area: C. 1816, there was 7.71 acres of arable land in four fields, and 396.18 of moorland, giving a total area of 403.89 acres.

Tenants: 1798, William Corbet and two others (71**).** Sold to Alexander Macdonald of Glenalladale in 1813 (49).

Settlements: NM 6860 7266. *Port a' Bhata.*
NM 6929 7197. *'Allt a' Mhuillinn'.*

Shielings: Not recorded.

Placenames:
1. **Portuvat (3). NM 6860 7308.** G. *Port a' Bhàta* 'the landing place of the boat', with gen. sg. of *bàta* m. 'boat'. The loch drains to leave a considerable expanse of soft mud, and the *port* can only be approached by small boats at flood tide. Larger vessels used *Camas Luinge* on the western side of the township.

2. **Ru Ard Duighnish (3). NM 6817 7304.** G. *Rubha Àird 'Duighnish'?* 'the point of the headland of Duighnish', perhaps with 'nish' a loan-word from Old Norse *nes,* 'headland'.

3. **Torramore (3). NM 6830 7257.** G. *An Tòrr Mòr* 'the large hill'. Named *Tòrr Port a' Bhàta* 'the hill of *Port a' Bhàta'* on OS maps (2). See no. 1, above.

4. **Ru an taul Donald (3). NM 6889 7250.** G. *Rubha Sabhal Dhòmhnaill* 'the point of *Dhòmhnaill*'s barn'. If the name does contain *sabhal* m. 'barn', one of two name forms are formally likely here: G. *Rubha an t-Sabhail* 'the point of the barn', with gen. sg. of *sabhal,* and G. *Rubha Sabhal Dhòmhnaill* 'the point of *Dòmhnall's* barn', with gen. sg. of the man's name, *Dòmhnall* 'Donald'.

5. **Cruach Bhach Dhuibh, 747 ft (4). NM 6815 7153.** G. Possibly *Cruach a' Bhaic Dhuibh* 'the hill of the black hollow ?'. Named *Beinn Bhreac, 240m* 'the speckled mountain' on OS maps.

6. **Allt Camas Longart (4). NM 6830 7240.** G. *Allt Camas an Longphuirt* 'the burn of *Camas an Longphuirt',* see Briaig no. 2'.

7. **Cnoc a Bhlair Cheann (4). NM 6866 7243.** The Gaelic name is obscure. Possibly G. *Cnoc a' Bhlair Cheann* 'the hillock of the end-plain ?'.

8. **Beinn a' Mhinn (16). NM 668 722.** G. *Beinn a' Mhinn* 'the mountain of the kid', with gen. sg. of *meann* m. 'kid', see Scardoish no. 5.

9. **Cnoc Longart (4). NM 6834 7291.** G. *Cnoc an Longphuirt* 'the hill of the harbour', with gen, sg. of *longphort* m. 'harbour'.

10. **Rudha Longart (4). NM 6824 7312.** G. *Rubha an Longphuirt* 'the point of the harbour', with gen. sg. of *longphort* m. 'harbour'.

11. **Allt na Cloich Dubh (4). NM 6854 7274.** G. *Allt na Cloiche Duibhe* 'the burn of the black stone', with gen. sg. of *clach.* f. 'stone, rock'.

12. **Lamraig na Baintighearna (16). NM 6807 7250.** G. *Lamraig na Baintighearn* 'the landing place of the lady', with gen. sg. of *baintighearna* f. 'lady'. The boat landing in *Camas Luinga*. Dwelly gives *Làimhrig Baintighearna* 'an easy landing place' i.e. where a lady could disembark easily.

13. **Camas Luinga (19). NM 6807 7250.** G. *Camas Luinga* 'the bay of the ships'. In May 1795, John Harriman of the Lorne Furnace Company wrote to Colin MacDonald of 'Dalela' *'the compy will need Grass for six or eight horses in or convenient to the woods of Dalela and Druimanloe to lead the coals, bark etc. to Loch Shield, and likewise they will need grass for twenty or twenty five Horses convenient to Carry over the Coals and Bark from Langall to Portvati from here'* (Ms. 995 f. 42v). He wrote again later on, complaining vehimently about how badly the horses had been treated in Moidart. The charcoal and bark were dispatched from *Camas Luinga*.

14. **Camas Longurt (3). NM 685 725.** G. *Camas an Longphuirt* 'the bay of the harbour'. See Briaig no. 2.

15. **Dundallan (34). NM 6816 7152.** Named *Beinn Bhreac* on OS maps.

Meall an Aoil = Hillock of Lime

16. **Mullen Ull (49). NM 6929 7198.** 'the mill built with lime'. Although built on Port a' Bhàta land, the mill appears to have been regarded as belonging to Langal. See Langal No. 17.

Lamraig na Baintighearn

186

Port a' Bhata: the 'port' from the W

The *port* after which the township was named in the lee of the headland   © Dr. Sandra Evans

ANNAT : An Annaid : the church.
1686, Annas (57); 1718, Annat (6); 1748, Annat (28); 1782, Annat (64); 1790, Annat (48); 1823, Annat (7).

Annat Fank and Rubha a' Chladha

Valuation: A ½ merk land in 1718, a ½ penny land in 1748, and a half merk land in 1790.

Area: C. 1816 there was one small field measuring 1.75 acres and 725.80 acres of moorland, a total of 727.55 acres, although there is an extensive area of lazy-beds on the hillside above the settlement.

Tenants: 1686, Effrick nein Donnochie, spouse to Ruary McInnes VcEwin (57); 1718, Donald McRory (6); 1744, Rory McVarish and John his brother (38); 1745, John Bain, John MacVorrich (12); 1748, Donald McVarish (32); 1782, Donald MacVarish (MacDonald) (64); Set in tack to Alexander Macdonald of Dalelia in 1790, including the Island Finan Ferry and Change house there, for 31 years (48). Sold to Alexander Macdonald of Glenalladale in 1813 (49).

Settlements: NM 7916 7199. Annat. Sheepfank on site of earlier settlement, with 6 quite complex structures shown on the estate plan, mainly within an enclosing (cashel?) wall, and a scatter of creel houses in the immediate vicinity. NM 8003 7260. Inner Gaskan. House and byre.

Shielings: Not recorded.

Placenames:
1. **Annat (3). NM 7916 7200.** G. *An Annaid,* from Old Irish *andóit*, a 'Mother Church'. This was almost certainly an Early Christian monastery possessing the relics of a local saint, and serving a number of chapels in the area. Its relationship to the potentially more important religious settlement on *Eilean Fhianan* is uncertain. A Bishop MacDougall, presumably of the Dunollie family, is said to have been buried here, suggesting that the burial ground was still in use during the Late Mediaeval period. The estate plan of c.1816 shows the secular settlement as a group of six complex structures, now in a very ruinous state, although there are numerous other buildings scattered over a wide area, which do not appear on the plan.

   Could be Moidart MacDugall

2. **Ru Claodh (3, 13). NM 7922 7200.** G. *Rubha a' Chladha* 'the promontory of the graveyard', with gen. sg. of *cladh* m. 'graveyard'. A substantial stone and turf dyke to the SE of the fank may mark the limit of either the burial ground or the monastic settlement, but it also enclosed most of the early 19th century *clachan*.

3. **Inner Gaskan (3). NM 8003 7260.** G. *?Inbher Gasgain* 'the mouth of the fish-tailed estuary'. Named **Inner a Gaskan** on Ms 384/1 (5).

Rubha a' Chladha from the NE with house and byre bottom left

4. **Ault Einish (3). NM 7992 7300.** G. *?Allt Eineach* 'the burn of bounty or goodness'.

5. **Cruoch Garme (3). NM 7755 7270.** G. *Cruach na Gairm* 'the hill of the calling', with gen. sg. of *gairm* f. 'calling'. Possiby G. *Cruach Gorm* 'the green hill'.

6. **Maol Or Ballach (3). NM 7856 7320.** G. *Maol Òir a' Bhealaich* 'the golden hill of the pass', with gen. sg. of *bealach* m. 'pass'.

7. **Cruach na Curagh (3). NM 7906 7376.** G. *Cruach nan Caorach* 'the hill of the sheep', with gen. pl. of *caorach* f. 'sheep'.

8. **Cruoch na March (3); Cruoch na marach (34). NM 8010 7467.** G. *Cruach a' Mharaiche* 'the hill of the sailor' or of 'the Scurvy grass', with gen. sg. of *maraiche* m., or *Cruach nam Maraiche* 'the hill of the sailors', with gen. pl. of the specific. *Cochlearia officinalis* 'Skurvy grass' is a common shore plant, but it also grows on wet ledges on the mountains.

9. **Illans Dou Gaskan (5). NM 8036 7248.** G. *Eileanan Dubha Ghasgain* 'the black or worthless islands of Gasgan'. *An Gasgan* might have the sense of 'plateau', 'copse', or even having the shape of 'a fish's tail'. Named *Eileanan Comhlach* 'the companion islands' on OS maps (2). The larger of the two islands has a fine 19th century Scot's Pine plantation, a nice Juniper bush, an excellent corn-drying kiln , the fragmentary remains of a vitrified fort and a *port* on the SE side with a flagged floor.

10. **Ballach (34). NM 7780 7270.** G. *Bealach* 'the pass'. Thomson puts this name at *Fireach Dubh*, but the ground there is steep and dangerous, especially on the northern side. *Bealach Coir' an Fheòir*, 1 km ESE and *Bealach Bàn*, 1.5 km to the NE, are far safer and both give good access between Annat and Glen Forslan.

LOCHANS : Na Lochanan : the little lochans
1686, Lochannan (57); 1718, Lochanns (6); 1739, Lochanan (1); 1749, Lochan (32); 1773, Lochans (46); 1823, Lochans (7,34).

Loch nan Lochan from 'Lochans'

Valuation: A ½ merk land in 1718 and in 1748.

Tenants: 1686, Gilespig McIlespick (57); 1718, Donald McGillespick alias McQueen (or Donald McGillespuk alias McQuirri (61); 1746, Duncan MacIsaac, Bailie of Moidart, John MacDonald (12); 1748, John MacIsaac, Ground Officer for Clanranald and Archibald MacIsaac (12);1749, Archibald McIsaac and Alex. McDonald, each with a ¼ merk land (32). 1773, an extention of the tack to James Macdonald, Tacksman of Lochans for 13 more years (46); 1782, 1798, Donald MacDonald (64, 71).

Settlements: NM 7431 7238. Lochans
                  NM 7384 7184. Coire na Criche.

Shielings: Not recorded.

Placenames:
1. **Lochanan (1). NM 7431 7238.** G. *Na Lochanan* 'the little lochs', with pl. of *lochan* m. This township was traditionally let to Clanranald's Baillie or Ground Officer.

Loch nan Lochan

Shepherd's House at Lochans with steep lazy beds on the hillside beyond

LEIDNACLOICH : Leathad na Cloiche : the hillside of the stone

1699, Leducloich (58); 1700, Ledacloich (58); 1718, Laidnacluoh (6); 1718, Laidnacloich (61); 1739, Leidnacloich (1); 1749, Lednacloich (32); 1798, Setuacloich (71); 1823, Letnacloich (7); 1832, Lot na cloch (34); 2007, Glenmoidart (2).

Leithad na Cloiche from Lochans

Valuation: A ½ merk land in 1718 and a 2 farthing land in 1749 (32).

Tenants: 1694, Dugald MacDonald with his wife Moir NcVarrish and children Eun and Mairi (58); 1718, Donald McLean (6); 1746, Angus Maclean, Donald Maclean (12); 1748, Allan McDonald and John McLean, 1 farthing each (32); 1798, Donald Macdonald (71).

Settlements: NM 7450 7275. Lednacloich.
                   NM 7458 7334

Shielings: Not recorded.

Placenames:

1.   **Leidnacloich (1). NM 7450 7275.** G. *Leathad na Cloiche* 'the hillside of the stone', with gen. sg. of *clach* f. 'stone'.

The sheepfank and a croft in the plantation to the right

191

GLENUIG: Gleann Ùige : the valley of Ùig (the bay)

1718, Glenving (6, 61); 1739, Glenuig (1); 1798, Glening (71); 1823, Glenuig (7); 2002, Glen Uig (37).

Two ports in the bay from the south-east

Valuation: A 1½ merk land in 1718 and a 6 farthing land in 1749 and 1798. There is a discrepancy here as in the Judicial Rental of 1718, John McInnish *depones the three penny part of the town, being a twenty shilling land is waste*' (61).

Area: c.1816, 93.1 acres arable, 1114.9 pasture and 4.5 lochs, total 1212.5 acres.

Tenants: 1718, John McCoil vic Neil, alias Burn, Donald McLean, John McInish alias McLean, Angus McLachlan, John McWilliam, Dugall McEwn and Christian McEwen, with ½m lying waste (6, 28); 1744, John McQueen, 1744 (38); 1746, John Maclean, Donald Maclean, John MacYonill vic Creul, Donald McDonald, ¾ farthing land, (12); 1748, Donald McIsaac, John Edmond, Donald McDonald, Angus McPherson and Mary McPherson each held ¾ of a farthing land, and Donald McInnes, Hugh Mcdonald, John McLean and Katherine McEachan each had ¼ of a farthing land (32); 1798, Angus Macdonald and twelve others (71). Feu disposition in favour of Major Allan Nicolson Mac donald dated 1827 (50).

Settlements: NM 669 767. *Glenuig.*

Shielings: NM 6855 7635. '*Loch Bealach na Gaoithe'.*
           NM 669 750. Close to *Aon Achadh na h-Àirigh.*

Placenames:
  1. **Glenuig (1). NM 669 767.** G. *Gleann Ùige* 'the valley of Ùig (the bay)', with a loan-name from Old Norse *Vik* 'bay'. The buildings were burned by the Hanoverian forces at the end of May, 1746.

  2. **Cnoc a Ceir (4). NM 6713 7508.** G. *Cnoc na Cèire* 'the hill of the buttock'; see no. 3, below.

  3. **Bealach a Ceir (4). NM 6696 7514.** G.Possibly *Bealach na Cèir* 'the pass of the buttock or haunch', with gen. sg. of *cèire* f. 'buttock, haunch, breech' if it does not have the sense 'candle'. Named *Bealach Caraich,* G. *Am Bealach Carach* 'the winding pass' on the 1:25,000 OS map and *Bealach Aon Achadh na h-Àiridh* 'the pass of the *Aon Achadh na h-Àiridh* (the shieling's only field)*'*, with gen. sg. of *àiridh* f, 'shieling' on the 1:10,560 scale maps. Ronald Black (2001, p.440) gives *cèir* 'wax' as a kenning for the white buttocks of the deer', so the name may be freely interpreted as 'the hill of the fleeing deer'. See Kylesmore No. 14.

  4. **Aird na Monadh (24). NM 671 771 ?** G. *Àird na Mònadh* 'the headland of the peat', with gen. sg. of *mòine* f. 'peat'.

192

5. **Ath na Moine (25).** G. Àth na Mòine 'the ford of the peat' with gen. sg. of *mòine* f. 'peat'. This is where the burn was crossed as peats were carried down off the hill to the village.

6. **Ballachan (24). NM 673 773.** G. Either *Ballachan* 'the speckled place' or *Bealachan* 'the small pass'. It is now used as a house name in the glen.

7. **An Dùn (25). NM 6714 7739.** G. *An Dùn* 'the hill or fort' with *dùn* m. but here *An Dùn* ''a lump (of rock) or heap'. This was the 'Parliament' place where Glenuig folk would meet in the morning to discuss the day's tasks.

8. **Port na Gralloch (25). NM 665 780.** G. *Port na Greallaiche* 'the landing place of the entrails', with gen. sg. of *greallach* f. 'entrails, gralloch'. Used by locals in which to kill and gralloch cows, a place which was away from the eyes of children, before they were salted and loaded onto boats for the market.

9. **Port nan Luinge (AM). NM 665 777.** G. *Port na Luinge* 'the landing place of the ship', with gen. sg. of *long* f. 'ship'.

10. **Sgeir Reithe Mòr (25). NM 673 778.** G. *Sgeir Reidh Mòr* 'the big smooth skerry'. Possibly G. *Sgeir an Reithe Mòr* 'the greater *Sgeir an Reithe* (the skerry of the ram)', with gen. sg. of *reithe* m. 'ram'.

11. **Sgeir Reithe Beag (25). NM 673 777.** G. *Sgeir Reidh Beag* 'the small smooth skerry'. Possibly G. *Sgeir an Reithe Beag* 'the lesser *Sgeir an Reithe* (the skerry of the ram)' with gen. sg. of *reithe* m., 'ram, tup'.

12. **Creag an Eighinn (26). NM 670 774.** G. *Creag an Eidhinn* 'the rock of the ivy', with gen. sg. of *eigheann,* here m., 'ivy'.

'Sadil', Glenuig Bay'.

13. **Sadil (34). NM 6726 7784.** G.? *An Sàideal* 'Bay'. Named Glenuig Bay on OS maps. *Sàideal* is an occasional place-name generic in Moidart and always refers to a bay. It does not appear to have been used further south, nor does it appear in 'Dwelly's Gaelic Dictionary'.

14. **Channa Cruoch (34). NM 6819 7612.** G. G. Possibly *Ceann na Cruach* 'the end of the hill', with gen. sg. of *cruach* f. 'hill or stack'. Named Glenuig hill on OS maps.

15. *Loch an Corrie Ily (34).* **NM 6820 7743.** G. Perhaps *Loch a' Choire eile* 'the other *Loch a' Choire* (loch of the corrie)', with gen. sg. of *coire* m. 'corrie'. Named *Loch a' Chairn Mhoir* 'the loch of the great cairn' on OS maps.

16. *Lochan Carlach (34).* **NM 6929 7582.** Possibly G. *Lochan a' Chàrlaich* 'the loch of the cartload', with gen. of *càrlach* m. 'cartload'. Possibly *Lochan na Cloiche Sgoilt* 'the lochan of the split stone' on OS maps.

17. **Seannlac (24). NM 6700 7612.** G. *Seannlag* 'the old hollow'. 'Seannlac' is the modern spelling of a house/croft name, but *Seannlag* is the correct Gaelic.

EGNAIG : Èigneig: Oak –neck (?): Oak bay

1718, Egnig (6); 1718, Egnaig (61); 1739, Eagneag (1); 1744, Eagnaig (38); 1749, Egnaig (32); 1782, Egneig (64), Eignoch, Eignaig (65); 1798, Equog (71); 1823, Egnaig (7); 2002, Egnaig (2).

A house on Egnaig
© Dr. Sandra Evans

Valuation: Egnaig was combined with Smirisary to form a 1 merk land in 1718. In 1748 it was a '5 chlitick or 1 farthing and ¼ farthing land, and a pertinant of the said town of Smirisary' (32).

Area: c.1816, 264 acres in total, with 18.0 arable and 246 pasture.

Tenants: 1718, John McCoil vic Inish (1/16m) and Angus McCoul (5/16m) (6); 1744, Angus McLean and Mary McDonald his wife (38); 1746, Donald Maclean, John Maclean (12); 1748, Angus Maclean (1¼ farthing land) (12, 32); 1752-54, Alasdair Mac Mhaighstir Alasdair (79); Angus MacDonald (64), Angus Beaton (65; 1798, Donald Macdonald (4/5) and John Macdonald (1/5). Feu Disposition in favour of Major Allan Mac donald dated 1827 (50).

Settlements:
NM 6593 7472. *Egnaig.* 13 structures on OS maps
NM 6660 7440. *Aultigil.* 5 structures.
NM 6540 7498. *Bad an Dobhrain.* At least two structures.
NM 6640 7454. *Saideal na h-Àirde Mòr.* 3 structures.

Shielings: Not recorded.

Placenames:

1. **Egnaig (2). NM 6593 7472.** G. *Èigneig,* probably a loan fron Old Norse; possibly *Eik-nakki* 'oak-neck', descriptive of a ridge, but the Gaelic ...*aig* may be ON *vik* 'bay'. There is a small unenclosed burial ground in close proximity to the settlement.

2. **Cnoc na Lon Ban (4). NM 6585 7560.** G. *Cnoc an Lòin Bhàin,* 'the hill of the white marsh', with gen. sg. of *lòn* m. 'meadow; brook; marsh', but in this location it is likely to be a marsh.

3. **Rudha Tigh Cladach (4). NM 6590 7465.** G. *Rubha Taigh a' Chladaich* 'the promontory of the house of the shore', with gen. sg. of *cladach* m. 'shore'.

4. **Camas Duilich (4). NM 6605 7460.** G. *An Camas Duilich* 'leaf bay'.

5. **Cnoc na Fang Bheag (4). NM 6567 7477.** G. *Cnoc na Fainge Bige* 'the hill of the little fank' or *Cnoc na Fainge Bheag* 'the lesser *Cnoc na Fainge'*, with gen. sg. of *faing* f. 'fank'.

Egnaig burial Ground   © Dr. Sandra Evans

6.   **Goirtean na Gobhar (4). NM 6564 7500.** G. *Goirtean nan Gobhar* 'the field of the goats', with gen. pl. of *gobhar* m. 'goat'.

7.   **Allt na Chu (4). NM 6549 7520.** Although the source form lacks an *–n,* it may represent G. *Allt nan Con* 'the burn of the dogs' or *Allt a' Choin* 'the burn of the dog', with gen. pl. or gen. sg. of *cù* m. 'dog', as inflected forms are sometimes restored to radical forms in borrowing.

8.   **Meall Mòr, 469ft. (4). NM 6511 7520.** G. *Am Meall Mòr* 'the large hill'.

9.   **Rudha na Caillich (4). NM 6628 7440.** G. *Rubha na Cailliche* 'the promontory of the old woman', with gen. sg. of *cailleach* f. 'old woman'.

10. **Sideal na h'Ard Mor (4). NM 6624 7448.** G.? *Sàideal na h-Àirde Mòire* 'the bay of the large headland'.

11. **Rudha nah Ard Mor (4). NM 6612 7447.** G. *Rubha na h-Àirde Mòire* 'the promontory of the large headland', with gen. sg. of *àird* f. 'headland'.

12. **Na Cathagh Ranhue (4). NM 6640 7521 (?)** An obscure name. A shoulder of Egnaig Hill, on the march with Glenuig.

13. **Cruachan na Muc, 740 ft (4). NM 6625 7528.** G. *Cruachan nam Muc* 'the hill of the pigs', with gen. pl. of *muc* f. 'pig', although *muc* here may be elliptical for *muc-fàileag* f. 'rose-hip'.

14. **Cnoc Mor na Leacann (4). NM 6580 7528.** Probably either G. *Cnoc Mòr na Leacainn* or *Cnoc Mòr nan Leacann* 'the large hill of the slope(s)', with gen. sg. or pl. of *leacann* f. 'slope'.

15. **Torr Buidhe (4). NM 6558 7530.** G. *An Tòrr Buidhe* 'the yellow hill'.

16. **Lochan Beamas (4). NM 6570 7590.** An obscure name. Named *Loch na Bairness* on OS maps.

17. **Allt a' Guil (4). NM 6679 7439.** G. Possibly *Allt a' Ghuil* 'the burn of weeping', with gen. sg. of *gul* m. 'weeping', or G. *Allt Gil* 'the burn of the ravine', with ON *gil* 'ravine'.

18. **Sron Mor (4). NM 6654 7461.** G. *Àn t-Sròn Mhòr* 'the large point'. Literally 'the big nose'.

19. **Ault Luglongurst (3). NM 6675 7433.** G. *Allt 'lug' Longphuist* 'the burn of the L…? harbour'. Possibly with *luchd* 'cargo'.

A house and byre at *Bad an Dobhrain*   © Dr. Sandra Evans

Aultigill

SAMALAMAN : ON Sam-holminum : the joined island (?)

1718, Sandalman (6, 61); 1739, Samlaman (1); 1744, Smilman (38); 1748, Samulaman (32); 1748, Samulaman (28); 1786, Samalaman (41, 62); 1823, Samalaman (7); 2002, Samalaman (2).

Samalaman House

Valuation: A ½ merk land in 1718 and a ½ penny land in 1748. A 2 farthing land in 1786 (41).

Area: In 1816 the 331.5 acres were made up of 17.75 arable and 315.75 pasture

Tenants: 1718, John McDonald alias McEan vic Coull (6) ; 1744, Donald McDonald and his spouse McVarish, and Donald's brother Lauchlan (38); 1748, John MacDonald (12); 1748, Donald McDonald and Laughlan McDonald his brother (indisposed) and Sarah McEachan their mother who is old and infirm do possess equally among them the ½ penny land (28); Bishop Alexander Macdonald and his successors in office for 35 years from Whitsunday, 1786 (41). (Bishop Chisholm in 1798). Feu Disposition in favour of Major Allan Mac donald dated 1823 (50).

Settlements: NM 6605 7758. Samalaman. The estate plan shows 8 structures to the west of the burn and 4 to the east.

Shielings: NM 6580 7613. 'Cruach na Bairness'. Rectangular structure.

Placenames:

1. **Samalaman (2). NM 6605 7758.** A curious name, likely to be a loan from an Old Norse form in –*holmi* 'island', perhaps ON *Sam-holminum* dat. In the sense of 'the joined island'. The township would therefore appear to take it's name from the tidal 'Samalaman Island'.

2. **Aird an Iasgaidh (16). NM 6533 7805.** G. *Àird an Iasgaich* 'the headland of the fishing' or *Àird nan Iasgair* 'the point of the fishermen', with gen. sg. of *iasgach* m. 'fishing'. Now known as *Ardbeg* 'the small headland'.

3. **Laran Mòr (25). NM 657 774.** G. An obscure name. Possibly G. *Làren Mòr* 'the big expanse of ground'.

4. **Pairc Ruadh (25). NM 662 777.** G. A' *Phàirc Ruadh* 'the red park or field'.

5. **Taigh an Fhigheadair (19). NM 659 769?** G. *Taigh an Fhigheadair* 'the house of the weaver', with gen. sg. of *figheadair* m. 'weaver'.

6. **Port Ard an Iasgaich (16). NM 653 779.** G. *Port Àird an Iasgaich* 'the landing place of *Àird an Iasgaich*', see no. 2, above.

7. **Poll Beag (17). NM 652 779.** G. *Am Poll Beag* ' the little marshy meadow'.

8. **Rubha nan Erricorry (24). NM 657 779.** An obscure name. G. *Rubha nan* 'Erricorry'. Possibly with G. *Èirigh* 'rising, surging' and *coire* 'whirlpool' or G. *Feurach Coire* 'grassy corry'.

9. **Rubh' a' Phuill Bhig (25). NM 653 781.** G. *Rubha a' Phuill Bhig,* 'the promontory of *Am Poll* Beag', see no. 7, above.

10. **Allt a' Gharbh Phuirt (16). NM 6600 7756.** G. *Allt a' Gharbh Phort,* 'the burn of *An Garbh Phort*'; 'the burn of the rough landing place'.

11. **Garbh Port (16). NM 6579 7790.** G. *An Garbh Phort* 'the rough landing place'.

12. **Ardennachan (21; 23). NM 6650 7802.** Possibly G. *Àird an Eanachaidh* ' the promontory of the Mat-grass (*Nardus stricta*)'. Named *Rubha na h-Àird Eanachaidh* 'the point of the stalky grass promontory' on OS maps. Wendy Wood and Margaret Leigh both referred to the headland by this name, so it must have been in the oral tradition of the district c. 1940.

Cruach Samalaman

13. **Cruoch Samalaman (34). NM 6600 7640.** G. *Cruach Samalaman* 'the hill of Samalaman'. Named *Cruach na Bairness* on OS maps.

14. **Saideal Samalaman (24). NM 6630 7785.** 'Samalaman Bay'.

Samalaman Bay and Island

SMIRISARY : Smearasairigh : the shieling of the butter

1700, Smirisary (58); 1718, Smirisaray (6, 61); 1749, Smirifary (32); 1827, Smerisary (50); 1832, Smerrassarry (34); 2002, Smirisary (2).

A croft on Smirisary

Smirisary

Valuation: Combined with Egnaig to form a 1 merk land in 1718. A 3 farthing land in 1749 (32).

Tenants: 1699, Angus oig N'Suine with Christian N'Kinvine his wife and Angus and Donald, their children; 1699, John McDonald, his wife Mary and children Duncan, Donald, Eun and Mary (58); 1718, John McCoil vic Inish (1/16m), Angus McEan vic Coill (3/16m), Niel McDonald (1/16m), Dougall McEun alias McDugall (⅛m) and Donald McGillichallum, Boat Wright or Carpenter to Clanranald, (⅛m), with ⅛m lying waste (6). 1745, John Mac Isaak, Angus Mac Isaak and Dugald MacDonald (22); Angus Mor McDonald, post 1745 (12); 1748, Dugald MacDonal, 1½ farthing land, John and Angus MacIsaac, 1½ farthing land (held equally) (12).

199

In 1748, John and Angus McIsaac possessed equally, 1½ farthing land, but John lived rent free as the Ground Officer. Another John McIsaac possessed a two chlitech or ½ farthings and Alexander McDonald and Christian McIsaac also possessed a ½ farthing land each (32). 1798, Donald Macdonald and seven others (71). Feu disposition in favour of Major Allan Mac donald dated 1827 (50).

Settlements: NM 6490 7709. *Smirisary.* A scattered crofting township with numerous structures spread over a wide area.

Shielings: NM 6540 7498. Bad an Dobhrain. House and byre.

Placenames:

1. **Smirisary (2). NM 6490 7709.** 'G. *Smearasairigh,* apparently from Old Norse *Smjørsærgi* 'the shieling of the butter', with gen. sg. of *smjør* nt 'butter'. This may be a vaccary name, and not that of a shieling.

2. **Badan Dunan (4). NM 6540 7491.** Perhaps G. *Bad an Dùnain* 'the thicket of the hillock or fortlet', with *bad* 'place; thicket' and gen. sg. of *dùnan* m. 'hillock, fortlet'. Named *Bad an Dòbhrain* 'the spot of the otter' on OS maps.

3. **Allt Bad an Dunan (4). NM 6530 7500.** G. *Allt Bad an Dùnain* 'the burn of *Bad an Dùnain;* see no. 2, above.' or *Allt Bad an Dòbhrain* 'the burn of the otter place', which seems more likely.

4. **Gleann Dunan (4). NM 6522 7522.** G. *Gleann an Dùnain* 'the glen of the hillock or fortlet', with gen. sg. of *dùnan* m. 'hilloch, fortlet', or G. *Gleann Dòbhrain* 'the glen of the otter'.

5. **Torr an Dunan (4). NM 6522 7565.** G. *Tòrr an Dùnain* 'the hill of the hillock or fortlet', with gen. sg. of *dùnan* m. 'hillock, fortlet', or G. *Torr an Dòbhrain* 'the hillock of the otter'.

6. **Cnoc Carn na Muc (4). NM 6486 7569.** G. *Cnoc Càrn nam Muc* 'the hill of *Càrn nam Muc* (the cairn of the pigs)', with gen. pl. of *muc* f. 'pig', although *muc* here may be elliptical for *muc-fàileag* f. 'rose-hip'.

7. **Cnoc Rirartan (4). NM 6472 7546.** The meaning is obscure. Possibly with *righe* 'shieling' and *artan* 'small stone, pebble'.

8. **Cruach Ghamaigh (4). NM 6491 7634.** G. ?*Cruach Ghamaigh* 'Gamie's (Gamekeeper's) hill'.

9. **Torran (4). NM 6465 7661.** G. *Na Torran* 'the hills'.

Cnoc na Beurrach or Smirisary Hill

10. **Cnoc na Beurrach (4). NM 6464 7596.** Possibly for G. *Cnoc a' Bhuirich* 'the hill of the bellowing/roaring'.

Crofts on Smirisary © Dr. Sandra Evans

11. **Allt Carn na Muc (4). NM 6483 7525.** G. *Allt Càrn nam Muc* 'the burn of *Càrn nam Muc* (the cairn of the pigs)'; see no. 6, above. The burn is not shown on OS maps but the contours on the 1:10,000 maps suggest one here.

12. **North Harbour (4). NM 6300 7600.** 'North Channel' on OS maps.

13. **Tobar Mhairearaid (6). NM 647 776.** G. *Tobar Mhairearaid* '*Mairearad's* well', with gen. sg. of the woman's name *Mairearad* 'Margaret'.

14. **Bogh Stree (4). NM 6404 7528.** G. *Bodha Strì* 'struggle reef', possibly referring to difficult tidal currents.

15. **Toll Uaine (6). NM 647 776.** G. *An Toll Uaine* 'the green hole'. A small byre by the sea; last lived in by Ceit Mhòr.

16. **I na Hotrach (4). NM 6435 7520.** G Named *Eilean na h-Oitire* 'the island of the low promontory jutting into the sea' on OS maps, which describes it well. Possibly G. *Eilean na h-'Otrach* ' the island of the midden', with gen. sg. of *òtrach* f. 'midden'.

17. **Ru Smerrassary (4). NM 6400 7546.** G. *Rubha Smearasairigh* 'the promontory of *Smearasairigh*'; see no. 1, above. Named *Rubha na Clach Dearga* 'the point of the red rock' on OS maps.

201

18. **Clach Dearg (4). NM 6400 7546.** G. *A' Clach Dhearg* 'the red rock'. An outcrop of hornblend-porphyry gives a reddish colour to the rock here.

19. **Bealach Rudha (27). NM 649 771.** G. *Am Bealach Ruadh* 'the red pass'.

20. **Cnoc na Bathaigh Mhòr (27). NM 649 771.** G. *Cnoc na Bàthaich Mòire* 'the hill of the big byre', with gen. sg. of *bàthach* f. 'byre'.

21. **Garbh Cheann Dubh (27). NM 649 773.** G. *An Garbh Cheann Dubh* 'the black *Garbh Cheann* (rough headland)'. The house at this point was occupied by Margaret Leigh, who wrote 'Spade Among the Rushes' during WW II.

22. **Goirtean (25). NM 648 774.** G. *An Goirtean* 'the field'. Katy Maclean (nee Gilles) lived here till Spring 1946, then it passed to Margaret Leigh.

23. **Larach an Tigh Duibh (27). NM 646 769.** G. *Làrach an Taigh-dhuibh* 'the site of the black-house', with gen. sg. of *taigh-dubh* m. 'black-house'.

24. **Lurmachd (27). NM 645 772.** G.? *Lurmachd* 'the exposed house', from *luramachd/lomnochd* 'naked, bare'?

25. **Pullag (21). NM 648 773.** G. Possibly from an obsolete word, *Pullag* 'pantry'. This was described by Margaret Leigh as a deserted cottage (where perhaps visitors were always well fed when it was still occupied?).

26. **Cnoc Mam a' Churaich (24). NM 650 773.** G. *Cnoc Màm a' Churaich* 'the hillock (or pass) of *Màm a' Churaich* (the hill of the coracle or marsh pass)', with gen. sg. of *curach* m. 'coracle, marsh'.

27. **An Druim (25). NM 653 772.** G. *An Druim* 'the ridge'.

28. **Faing Mhic Phaill (25). NM 650 773.** G. *Faing MhicPhàil* '*MacPhàil's* fank', with gen. sg. of *MacPhàil* 'MacPhail'. A sheepfank named after a man called MacPhail, who died there suddenly when gathering.

29. **Goirtean na Gobhar (4). NM 653 752.** G. *Goirtean nan Gobhar* 'the field of the goats', with gen. pl. of *gobhar* m. 'goat'.

30. **Blàr Eorna (25). NM 652 774.** G. *Am Blàr Eòrna* 'the barley field', with gen. sg. of *eòrna* m. 'barley'. The croft house here was tenanted by Donald MacIssac during WW II.

31. **Blaran Boidheach (25). NM 648 774.** G. *Am Blàran Bòidheach* 'the pretty field'. A field of about 1.25 acres lying next to the house occupied by Margaret Leigh, who described it as 'Of irregular shape, fenced on three sides by a low, casually arranged drystone dyke and the fourth side a sheer drop to the foreshore, with the remains of a house, byre and three other structures'.

32. **An Lòn, (25). NM 646 773.** G. *An Lòn* 'the meadow'. Cultivated by the Smirissary crofters until 1950.

33. **Am Bogha Dearg (25).** NM 640 755. G. *Am Bodha Dearg* 'the red submerged rock'. Sometimes referred to as *Am Bò Dearg* 'the red cow'.

34. **Red Cow (4).** NM 642 772. G. *Am Bò Dearg* 'the red cow', but as the name comes from the Admiralty Chart, *Am Bodha Dearg* seems more likely. See no. 33, above.

35. **Eilean na Fiach (4).** NM 6435 7518. G. Possibly *Eilean nan Fitheach* 'the island of the ravens', with gen. pl. of *fitheach* m. 'raven'. Named *Eilean na h-Oitire* 'the island of the low promontory jutting out into the sea' on OS maps.

36. **Laran Mòr (21).** The meaning is obscure. G. *?Làren Mòr* 'the big expanse of ground'. This is where Margaret Leigh used to cut her peats.

37. **Achadh an Aonaich (21).** NM 6445 7645 . G. *Achadh an Aonaich* 'the field of the hill', with gen. sg. of *aonach* m. 'hill'. An area of irregular lazy-beds amongst rocky outcrops.

38. **Glac nan Lion (21).** G. *Glac nan Lìon* 'the hollow of the nets', with gen. pl. of *lìon* m. 'net'.

39. **Fang Mòr (21).** NM 6457 7747. G. *An Fhaing Mhòr* 'the big fank' or 'the great fold'.

40. **Lòn Liath (21).** G. *An Lòn Liath* 'the grey (hoary) marsh'.

41. **Cnoc a' Mhullain (21).** G. *Cnoc a' Mhullain* 'the hill of the mole', with gen. sg. of *mullan* m. 'mole', or perhaps G. *Cnoc a' Mhulain* 'the hill of the stack', with gen. sg. of *mullan* m. 'corn stack'. *Cnoc a' Muilinn* 'hill of the mill' is also possible.

42. **Cnoc an t-Sabhail (21).** G. *Cnoc an t-Sabhail* 'the hill of the barn'.

43. **Tigh Lachain (21).** G. *Taigh Lachainn* '*Lachann*'s house', with gen. sg. of *Lachann,* an occasional form of the man's name *Lachlann* 'Lachie'.

44. **Glac Cnuic an t-Sabhail (21).** G. *Glac Cnoc an t-Sabhail* 'the hollow of *Cnoc an t-Sabhail*', see no. 42, above.

45. **Sean Rudha (25).** NM 64 77. G. *An Seann Rubha* 'the old promontory'.

46. **Am Bùth (27).** NM 646 771. G. *Am Bùth* 'the bothy or shop'.

47. **Pol Beag (33).** G. *Am Poll Beag* 'the small pool', overhung by a cliff.

48. **Glac Nighiderachd (33).** G. *Glac na Nigheadaireachd* 'the hollow of the washing', with gen sg. of *nigheadaireachd* f. 'washing'. Described as the 'place up the quarry beside the stream where the washing, and particularly the trampling of blankets, used to be done'.

49. **Blair Geamhraidh (33).** G. *Am Blàr Geamhraidh* 'the winter field'. A sheltered place where cattle grazed in winter.

50. **Làren Beag (33).** The meaning is obscure. Possibly G. *? Làren Beag* 'the small expanse of ground'.

51. **Parc Dubh (33).** G. *A' Phàirc Dhubh* 'the black field'.

52. **Creag Eidheann (33).** G. *Creag an Eidheann* 'the rock of the ivy', with gen. sg. of *eidheann* m. 'ivy'.

53. **Port Luinge (33).** G. *Port na Luinge* 'the landing place of the ship', with gen. sg. of *long* f. 'ship'. A sailing ship was once wrecked here.

54. **Larach a' Bothan (33).** G. *Làrach a' Bhothain* (singular) or *nam Bothan* (plural) 'the site of the bothy', with gen. sg. of *bothan* m. 'bothy' or *nam Bothan* 'bothies'. The tinkers used to stay here when they were travelling through the district.

55. **Garbh Port (33).** G. *An Garbh Phort* 'the rough landing place'.

56. **Bidean Biorach (33).** G. *Am Bidean Biorach* 'the pointed pinacle'.

57. **Gead a' Leigich (33).** G. *Gead an Lighiche* 'the doctor's rigg', with gen. sg. of *lighiche* m. 'doctor, physician'.

58. Feadan Ban (33). G. *Am Feadan Bàn* 'the white gully'. Wendy Wood wrote that this 'is the rock funnel we used when we wanted to drop down to our neighbour's house, for in a wind it whistles the weirdest tunes', suggesting that the name could translate as 'the white chanter', but goes on to say that it could also be the 'White Waterfall', which it is after heavy rain.

59. Sean Bhaile (33). G. *An Sean Bhaile* 'the old village' or 'old homestead'.

60. Ru Chaidachleach (34). NM 6493 7789. G. Possibly *Rubha Gead an Lighiche* 'the promontory of *Gead an Lighiche'*; see no. 57, above. Named *Rubha Ghead a' Leighe* on OS maps.

61. Ru Port Bulogh (34). NM 6460 7754. G. Possibly *Rubha Port a' Bhealaich* 'the promontory of the landing place of the pass'. Cf. no. 19, above. Possible named *Rubha na Faing Mòire* 'the promontory of the big fank' on OS maps.

Eilean Coille with *An Glas-eilean* behind

62. Culagh Is (34). NM 6382 7627. G. *Eilean na Cullach* 'the year old calf island', i.e. seen to be a calf to the mainland. Possibly an angicisation of G. *Eilean Coillteach* 'woody island' as given on the 1st Edition OS map. Named *Eilean Coille* on later OS maps, but it is too low and exposed to have ever carried woodland.

One of the caves used for storage

LANDS IN MOIDART PERTAINING TO DONALD MACDONALD OF KINLOCHMOIDART, 1739.

UPPER KINLOCH : Ceann Loch Uachdarach : upper Ceann Loch

1695, Kinlochanquoich (58); 1739, Upper Kenloch (1); 1748, Kenlochmore (28);1789, Kenlochuachkerach (63); 1802, Craig (70); 1823, Kinlochmoidart (7); 1845, Kinloch (9).

Valuation: Kinlochmore, Kinlochtera, Badnacraggan and Brunary formed a 3 merk land in 1748. Kinloch was a one merk land in 1845 (8), but this was probably copying from an earlier document. In 1749 a 1 merk land was the equivalent of 13s 4d land of old extent in Moidart, but a 2d land in Sunart.

Tenants: 1695, Duncan McCarmaig, his wife Margaret McLain and their children John and Donald (58); 744, Mr Walter Johnstone, Schoolmaster (38); 1746, John MacFinla vic Ean Roy, Ian Macyonill vic Ean Yonill Oig, Donald MacPherson, Angus MacPherson, John MacInnes vic Ean vic Creul, Donald MacMyllan, Ranald MacDonald, two John MacDonalds, and John MacIsaac the Violer (12); 1749, Dr. John MacDonald, 1 merk, Alexr. McDonald, 5/-, Anne McDonald 6/8d, Duncan Boyd, 1/8d, John McEacharan, 3/4d, Neil Beaton, 1/8d, Donald McPherson, 1/8d, John Smith, 3/4d, John McPherson, 10d, Dougal Roy, 10d and Hugh McLeod, 1/8d of the 3 merk land (28); 1802, John MacDonald, tenant of Craig (70).

Settlements: NM 7180 7276.

Shielings: NM 7075 7474. *Airidh an Aon bhuinn.*
              NM 6917 7473. '*Loch Ard a' Phuill*'.

Placenames:

1. **Upper Kinloch (1). NM 7180 7276.** G. *Ceann Loch Uachdarach* 'Upper Ceann Loch (loch-end)'.

2. **Craigleabegg (28). NM 7113 7320 (?)** G. *Creag Liath Bheag* 'the lesser *Creag Liath* (grey crag)', 'the small grey crag'. The OS map puts *Creag Liath Bheag* at NM 7100 7388, but David Bruce gave this name to an area of woodland in 1748, and the OS position is too high and quite unsuited for an area of woodland which, together with the wood of Craig Kinloch, were valued at £15. The suggested position is likely to have been an ancient woodland site. By 1789, there was *only a little blackwood with out any oaks* (63). Used for summer grazing in the late 1700s (66).

3. **Craig Kenloch (28). NM 6929 7317 (?)** G. *Creag Cheann Loch,* 'the crag of *Ceann Loch*'. David Bruce mentions a wood of this name on the north side of Lochmoidart in 1748, and this is the most likely place.

4. **Loch Arrie Ghonal Ban (34). NM 7045 7613.** G. *Loch Àirigh Dhòmhnaill Bhàin* 'the loch of *Dòmhnall Bàn*'s shieling', with gen. sg. of *Dòmhnall Bàn* 'fair Donald'. Named *Lochan Domhnuill* 'Donald's lochan' on OS maps.

5. **Craig (66).** Townships appear to have been divided about the year 1800 and new names appear. 'Craig' may be part of Kinlochmoidart or Kylesbeg. In 1809, Craigs was set in tack to Alexander Gillies, Claish, Sunart, and Alexander Cameron, Ardtoe, but described as being 'at present in the possession of Mr. John Stewart (72).

6. **Kincarra (66). NM 7003 727280.** G. *Ceann na Carraigh* 'the end of the projecting rock', with *carragh* gen. *carraigh* 'projecting rock, pillar'. A small farm close to the site of the original Kinlochmoidart House.

Langal and Port a' Bhata from Upper Kinloch

NETHER KINLOCH : Ceann Loch Ìochdarach : lower Ceann Loch

1739, Neither Kenloch (1); 1744, Kenlochuachtrich (38); 1748, Kinlochuachair (12); 1748, Kenlochtera (28); 1802, Kenlochuachdrach (70); 1845, Kinlochachrioch (9).

Nether Kinloch from the west

Valuation: A 1 merk land with Kinlochmore and Badnacraggan in 1748. A merk land, together with Badnagrogan in 1845 (9).

Area:

Tenants: 1744, Mrs. Effy McDonald (38); 1748, Dr. John McDonald (12); 1802, Donald MacDonald, Lochans (70).

Settlements: NM 7204 7162. Lower Kinloch.

Shielings: Not recorded.

Placenames:
1. **Neither Kenloch (1). NM 7102 7161.** G. *Ceann Loch Ìochdarach* 'lower *Ceann Loch*'.

2. **Kinlochachrioch (9).** As above. G. ?*Cean Loch Achaidh Riabhach*

3. **Loch Uachdarach (26). NM 7050 7220.** G. *An Loch Uachdarach* 'the upper loch', 'the uppermost part of the loch'. This is the old name given to the upper (east) end of Loch Moidart.

4. **Kinlochuachair (12). NM 7050 7220.** G. Probably *Ceann Loch Ìochdarach* 'lower *Ceann Loch'*. This is the name given to the settlement in 1748.

A creel house on Nether Kinloch with the corners marked in red

Nether Kinloch from the west

Coffin cairns on the road to Dalelia and the Green Isle, with Loch Moidart in the distance

Nether Kinloch, with the settlement by the rocky outcrops, and peat banks in the moss in the foreground

BADNACROGGAN : Bad nan Crogan : the place or thicket of the jars
1739, Badnacroggan (1); 1749, Badnacraggan (12); 1789, Badnagrogan (63); 1802, Ardmolach (70); 1845, Badnagrogan (8).

Kinlochmoidart Bridge, built to a Thomas Telford design c. 1807

Valuation: A one merk land with Kinlochmore and Kinlochuachair in 1748 (12); A merk land together with Kinlochachrioch in 1845 (8). The wood of Badnacraggan was valued at £5 in 1748/9 (28), but it now appears on the map as 'Ardmolich Wood'.

Tenants: 1748, Donald MacPherson (12); 1802, Alexr. and Donald MacDonald (70).

Settlements: Possibly NM 7127 7199, Ardmolich.

Shielings: Not recorded.

Placenames:
1. **Badnacroggan (1).** G. Possibly *Bad nan Crogan* 'the place or thicket of the jars', with gen. pl. of *crogan* m., 'jar, pitcher, dish'. The specific could be a mis-spelling of *Creagan* 'little rock or crag'. The location of the settlement is not known but there are a number of small buildings scattered through the woods here. In 1789 it was recorded that '*upon Badnagrogan there are some straggling oak trees amongst a considerable extent of birch and other blackwood, that runs alongst the hill on the south side of the Strath of Moydart which upon the whole it is computed will yield no more bark than...2 tons*' (63). There are several charcoal-burner's pitsteads along the *Glaic Mhòr*.

2. **Tienganahimristan (66), NM 706 716.** Three small enclosures here were being used for grazing in the late 18th century. These have not been located but must be in the vicinity of the *Allt Teang' Emilstoin*. The Gaelic is uncertain but possibly with *imreasan* gen. *imreasain* 'dispute, controversy'.

3. **Cnoc Ardmolich (16). NM 710 721.** G. *Cnoc na h-Àirde Molaich* 'the hill of the wooded headland', with gen. sg. of *An Àird Mholach*.

KYLESMOR and KYLESBEG: An Caolas Mòr and An Caolas Beag : the greater and lesser (township of) Kyles

1744, Colis (38); 1746, Caolas (12); 1749, Kulis Ian Oig, Kulis More (28); 1782, Kylesmoidart (64); 1784/85, Kylesinog (67); 1786, Caolisbeg, Caolismoir (68); 1789, Kilismore; Kilisbeg or Kiliscolta (63); 1792, Killiemore, Killiebeg (69); 1809, Caolas Beg, Caolas Mor (72); 1845, Killescello (8); 1845, Chulsmoir (8); 2007, Kylesmor; Kylesbeg (2).

Ruined building in Kylesbeg Wood   © Dr. Sandra Evans

Valuation: Kulis Ian-oig, Kulismore and Shunavegg formed a 1 merk land in 1748 (28) with Shona Beag a ¼d land (8). Killescello and Chulsmoir were valued at ½ merk each in 1845.

Area: There is no obvious march dyke between the two townships on the OS maps, although it should be obvious on the ground. The hill grazing may well have been shared between them.

Tenants: 1744, John McEachin (38); 1746, John MacNeill Mor, Ruari MacInnish Mor (12); 1748, John McEacheran (Iain Òig), ¾ m and Angus MacPherson ¼m (28); 1782, Alexander MacPherson (64); 1784/85, Ewen, Alexander and John MacDonald, Angus and Niel Roy MacInnes (67); 1786, Donald MacIntyre (Ground Officer) in Caolisbeg; Ewen MacDonald, John, Dugald and Alexander MacPherson in Caolismoir (68); 1792, Killiebeg set in tack to Peter MacIntyre (Ground Officer) there, and Killiemore to Jas. MacDonald (69); 1802, Peter MacIntyre, tenant in Killesbeg and James MacDonald, tenant in Killesmore (70); 1809, both townships set in tack to Alexr. Gilles, Claish, and Alexr. Cameron, Ardtoe, but formerly set to Mr. John Stewart (72).

Settlements: NM 6706 7413. *Kylesmore.* A scattered group of at least 7 structure.
                         NM 6755 7393. *Kylesbeg.* A scattered group of at least 6 structures.
                         NM 6791 7373. '*Port an Dunan*'.

Shielings: NM 6913 7467. '*Lochan Ard a' Phuill*'.
                      NM 6800 7482. '*Coire Doir' Uillt*'. Developed shieling; two fields with prominent lazy-beds.

Placenames:
1. **Caolas (12).** G. *(An) Caolas* 'the strait' or 'the sound'. This appears to be the original name of the township, but at some time it was divided into two, *Caolas Mòr* and *Caolas Beag,* 'the greater/lesser *Caolas'.*

2. **Kylesmor (2). NM 6706 7413.** G. *(An) Caolas Mòr,* literally 'the large strait', 'greater *Caolas',* but in this context it means the 'larger division of *Caolas* township'.

3. **Kylesbeg (2). NM6755 7393.** G. *An Caolas Beag,* literally 'the small strait','lesser *Caolas',* but actually the 'smaller division of *Caolas* township'.

4. **Allt an Leanabh (4). NM 6713 7404.** G. *Allt an Leanaibh* 'the burn of the child', with gen. sg. of *leanabh* m. 'child'.

5. **Allt na Innes Chulun (4). NM 6750 7443.** G. *Allt na h-Innse Chuilinn* 'the burn of the holly meadow', or *Allt Innis a' Chuilinn* 'the burn of *Innis a' Chuilinn*', see no. 11, below. Named *Allt na h-Uraich, possibly Allt na h-Iubharach,'* the burn where yew trees (probably Juniper) abound', or *Allt na h-Ùraich* 'the burn of the soil' on OS maps.

6. **Loch Innes a Chulun (4). NM 6745 7493.** G. *Loch Innis a' Chuilinn* 'the loch of *Innis a' Chuilinn* (the holly meadow)', see no. 11, below. Named *Loch na Draipe* on the OS maps.

7. **Cona (4). NM 6871 7457.** Perhaps this is a miswriting or miscopying of *coille* or *coire,* or even *cona* f. 'Scots pine' or *connadh* m. 'firewood', but none of these suggestions seem plausible.

8. **Bealach an Dunan (4). NM 6828 7434.** G. *Bealach an Dùnain* 'the pass of the fort', with gen. sg. of *dùnan* m. 'fort, hill', although this appears to be a route to nowhere. Named *Bealach nan Coisichean* 'the pass of the walkers' on OS maps. The summit of the pass on the old road is very close to the entrance to the spectacularly sited fort at NM 6825 7363, which seems a more likely location for this name.

9. **Porst an Dunan (4). NM 6804 7357.** G. *Port an Dùnain* 'the landing place of the fort', with gen. sg. of *dùnan* m. 'fort'.

10. **Meall Mòr (4). NM 6820 7357.** G. *Am Meall Mòr ,*'the large hill'. Named *An Dùn* 'the fort' on OS maps.

11. **Innes a Chulun (4). NM 6774 7388.** G. *Innis a' Chuilinn* 'the meadow of the holly'.

12. **Cnoc an Luath, 711 ft (4). NM 6859 7410.** G. *Cnoc na Luaith/Luatha* 'the hill of the ash', with gen. sg. of *luath* f. 'ash'. Named *Cruach na Cuilidh Bige,* 238m, 'the hill of the small hollow' on OS maps.

13. **Kulis Ian Og (12). NM 6791 7373.** G. *Caolas Iain Òig ,* '*Iain* Òg's straits', with gen. sg. of the man's name *Iain Òg* 'Young Iain', but probably '*Iain Òg*'s portion of *Caolas* township', a colloquial name for *An Caolas Beag* 'Kylesbeg', or a portion of it, in 1748. *Iain Òg* may have been the by-name of John McEacheran who was the principal tenant at that time.

14. **Cnoc a' Ceir (4). NM 6712 7508.** G. *Cnoc na Cèire or Cnoc nan Cèir* 'the hillock of the wax or candle(s)', with gen. sg. or pl. of *cèir* f. 'wax, candle'. Ronald Black (2001 p.440) gives cèir 'wax' as a kenning for 'the white buttocks of the deer'. There is good sheltered grazing for deer below the pass, and when the deer were disturbed, all the travellers would see of them would be flashes of white bobbing over the skyline, so the name may be freely translated as 'the hill of the fleeing deer'.

An Dùn, crowned with an Iron Age fort

15. **Plate Rock (14). NM 6822 7357.** An alternative name for *An Dùn* 'the fort'. This is where MacDonald of Kinlochmoidart hid his silverware in an attempt to stop the Hanoverian forces from stealing it in May-July 1746.

16. **Eas Ailean (14). NM 6868 7341.** G. *Eas Ailein* 'Ailean's waterfall', with gen. sg. of the man's name *Ailean* 'Alan'. This is where the old track from Kinlochmoidart to Glenuig crosses the *Allt Ailein* 'Alan's burn'.

17. **The Ladder (14). NM 6867 7340.** Immediately to the west of the *Allt Ailein* the old road climbs very steeply up a remarkable zig-zag, a visionary piece of civil engineering. In the past it was much photographed as **The Devil's Staircase,** an exciting stage in Lochaber's famous 'Scottish Six Days Motor Cycle Trials'.

18. ***The old wood of Kulis (28). NM 6735 7395.*** This is how David Bruce described a wood on the north side of Loch Moidart in 1748, when it was valued at £2. It can only be 'Kyles Wood' below *Tòrr Mòr*. By 1789 *'On Kilismore there are no oaks except a few scatter'd over the face of a hill opposite to the island of Shunabeg that are computed to give bark of ..10 cwt'. 'On Kilisbeg or Kiliscolta...there is a thriving stool of oakwoods of considerable value'* (63).

Innis Chuilinn - Holly Dale.

Houses on Kylesbeg

© Dr. Sandra Evans

BRUNARY : Brunairigh : the spring/well shieling

1700, Brunarie (58); 1748, Brunary (28); 1789, Bronarie (63); 1845, Brouarie (8); 2002, Brunery (2).

Valuation: A one merk land in 1845.

Tenants: 1700, John McDonald, with Mary NcVarrish, relict and Donald, Eune, Dugald and Mary, their children, (58); Let with the 3 merk land of Kenlochmore, Kenlochochtera and Badnacraggan in 1748. See under 'Upper Kinloch'. 1809, John MacDonald, tenant (71).

Settlements: NM 7248 7197. Brunary.
                   NM 7394 7272.

Shielings: Not known.

Placenames:
1. **Brunery (2). NM 725 720.** This may represent either G. *Brunairigh* or G. *Brùnairigh.* If the former, it might be a loan from Old Norse *Brunn-ærgi* 'spring shieling', with the stem form of *brunnr* m. 'spring, well'. If the latter, it might be a loan from Old Norse *Brún(ar)ærgi* 'the shieling of the brow', with stem form or gen. sg. of *brún* f. 'brow of hill'.

Torr Cill Uanding

2. **Torvickluntin (28). The wood park of Torric-Killuntin (66). NM 7280 7200.** A hill name, but the meaning is obscure. David Bruce mentions a wood of this name on the north side of the Kinloch water and valued at £5 in 1748. In 1789 it was described as *'a close thriving stool of oakwoods of considerable value'* (63). Named 'Torr Cill Uanding' on OS maps. The name suggests an Early Christian site, a church or monastic settlement dedicated to St. Fhionntáin or St. Fintán. The *Martyrology of Donegal* lists 22 saints of this name, and little is known of any of them.

Lazy-beds on the brackeny knoll, with the march with Leidnacloich following the burn

ULGARY : Ulfhghearraidh : wolf-field (?)

1686, Ulogory (57); 1700, Ulgary (37); 1739, Ulln Corrie (1); 1749, Ulgory (28); 1845, Ullocarie (8); 1786, Ulagarry (68); 2002, Ulgary (2).

Ulgary from the SW   © Dr. Sandra Evans

Valuation: A one merk land in 1748 and in 1845. A 1d land in 1786 (68).

Tenants: 1686, Ewin McDougald VcInnes (57); 1698, John McDonald with Effie McDonald, relict and Donald and Katherine, their children (37); Laughlan McDonald, his wife Mary nin Ian vic Donill vic Vurchie and their children, John, Donald, Dugald and Anna (58); 1700, Mary, nin Ian vic Donell vic Vurchie, spouse to Lauchlan McDonald, and their children, John, Donald, Dugald and Anna (58); 1745, John Macintyre the piper to Kinlochmoidart; John Ban MacDonald, John MacDonald, Ewan Ban MacDonald in1746; his son Angus McDonald alias McEwan 1/3d, and grandson, Alexander McLachlan vic Ewan, 6/3d, were tenants there in 1749, along with John McDonald, 1/8d, Angus McDonald, 1/8d and Rory McDougal, 2/6d ; total 13/4d (28). 1786, Rory, John, Ewen Sr., Angus, Ewen Jnr., Donald and Dugald MacDonald, and John MacPherson (68); 1792, set in tack to Archibald MacDonald, Duchamis (69); 1802, John MacDonald, Borrodale, tenant (70); 1809, Peter Stewart, Glenforsland (71).

Settlements: NM 7735 7602. *Ulgary.* A scattered settlement with over 40 structures.
                   NM 7810 7614. Possibly shielings.

Shielings: Not recorded.

Placenames:
1. **Ulgary (1). NM 7735 7602.** Possibly Old Norse *Ulf-gerði* 'wolf-field' might yield G. *Ulfghearraidh,* which could be represented here by Ulgary, but it is not certain what the Gaelic pronunciation was. Ulgary appears to have been the abode of many great pipers. Long years ago, one Ulgary piper made a set of pipes, and the faerie suggested an extra hole in the chanter, promising that the result should be that, upon whatever occasion they were played, victory for Scotland would follow. It is known that they were played at Bannockburn (14). Alasdair Cameron suggested 'Bog Cotton shieling' while Tearlach MacFarlane accepts it as incorporating a personal name, 'Ulla's shieling'. 'Ulf's vaccary or shieling' is a possibility.

2. **Pole-a-bhainne (22). NM .** *Poll a' Bhainne* 'the hole of the milk'. '*The deer when visiting the crops (in Ulgary) used to come down from the high grounds through a narrow gorge with very steep sides, called Poll-a-bhainne, and the tenants, perfectly aware of their habits, by erecting a rude but effective fence across the lower end of the gorge, and then seizing the upper end and enclosing it in a similar manner after the deer had entered, kept them prisoners as effectively as any sheep in a fank. The tenants, ...by way of compensation, were at once assisted by the proprietor in having their arable ground well fenced*' (22).

214

Ulgary looking across to Doilet   © Dr. Sandra Evans

Ulgary looking west to Assary   © Dr. Sandra Evans

Ulgary with Doilet across the river © Dr. Sandra Evans

Ulgary © Dr. Sandra Evans

AORINN : Aoireann : foreshore
1718, Aorin (6); 1739, Aorinn (1); 1746, Irinn (12); 1748, Irine (28); 1782, Irin (64); 1823, Irine (7); 2002, Roshven (2).

Rois Bheinn and An Stac from the vitrified fort on Eilean nan Gobhar

Valuation: A 1 merk land in 1718 and in 1748, when it was wadset along with the ½ merk land of Forfey (28).

Tenants: 1718, held on a wadset by Ranald McDonald of Kenloch Moydart; 1746, John, Daniel, Donald and Duncan Cameron, Ewan MacVorrich and Donald MacNeill Mor (12); 1748, Ranald McDonald of Daliburgh was tenant of the 1 merk land of Irene and the ½ merk land of Forsay, and evicted by Clanranald in 1781. Tack by Clanranald and Trustees to Capt. Ranald McDonald for 13 years from Whitsunday 1821 (39). 1782, Dr. MacEachren / MacEachen, James MacDonald (64), 1782, Ronald MacDonald, later in Langale (65); 1798, set in lease to John Maceachan Snr. for 40 years from 1781 (71).

Settlements: NM 7190 7853. *Aorinn.*'Roshven'.
  NM 689 779.   *Forsay.* A field system but no remains of a settlement are known.
  NM 6790 7805. House and byre on the side of the old road.

Shielings: NM 699 763. 'Allt Mhic Eoghain'.
NM 7070 7603. *Airidh Domhnuill Bàn.*

Placenames:

1. **Aorinn (1). NM** G. *Aoireann,* f. 'foreshore, beach', *Irionn* 'the field', *na h-Aorainn* 'the place of worship' , *Earrann* 'a section of land' and a sandbank or shoal have also been suggested. It may also be cognate with *Aoirneagan* 'wallowing'. *correct – Iain MacMaster*

2. **Forsay (2), NM 6890 7798.** ON. *Fors-à* 'the waterfall burn'. There are spectacular waterfalls on the burn to the east of the settlement and another to the west.

3. **Coill' a' Chairn Mòr (16). NM 680 782.** G. *Coille a' Chàirn Mhòr* ' the wood of the large cairn', with gen. sg. of *càrn* m. 'cairn'.

4. **Lochan Donn (16). NM 7150 7600.** G. *An Lochan Donn* 'the brown loch'. **Loch Arrie Ghonal Ban (34).** *Loch Àirigh Dhòmhnuill Bhàn* 'the loch of *Dòmhnall Bàn*'s (Fair Donald's) shieling'.

5. **Ru Cairn Douran (34). NM 7005 7881.** Possibly G. *Rubha Càirn Dòmh'ain* 'the promontory of *Dòmh'an*'s cairn', with gen. sg. of *Dòmh'an,* a short form of the man's name *Dòmhnallan,* diminutive of *Dòmhnall* 'Donald'.

6. **Ru ard na Bay (34). NM 7054 7922.** G. *Rubha Àird a' Bhàigh* 'the promontory of the headland of the bay' or *Rubha Àird a' Bheith* 'the promontory of the headland of the birch', with gen. sg. of *bàgh* m. or *beith* m. 'birch'.

7. **Lachk Ferry (34). NM 6849 7812.** 'the ferry of the slabs'. This was Bonnie Prince Charlie's first landfall in Moidart in 1745.

A port near the outfall of the Irine Burn

An Stac on Easter Alisary from the Roshven Burial Ground

Inverness Sasines dated 1829 and 1850 mentions 'Lower Polness or Polnish and Laggan with the islands called Orassary (or Orosarry), Nabarnnach (or Nabarnoch) and Nacharish (or Nachrist) in Loch Ailort belonging thereto' (35). These townships form part of the 24 merkland of Arisaig, but as they are close to Aorinn they are included here for the sake of completeness.

8. **Orassary; Orosarry (35). Orrasary I. (34). NM 722 797.** A tidal island named on OS maps as *Eilean nan Trom*. The documentary form sounds like a shieling name, but it could be a corruption of Oronsay, a 'tidal island'. Perhaps G. *Orasaraigh*, rather than simply *Orasaigh*, a loan possibly from Old Norse *Orfirisøyjarærgi* 'the shieling of *Orfirisøy* (the tidal island)'. *Eilean nan Trom* 'the island of the goods or cargo'. There is deep water on the south side and the mail boat could tie up and unload here at any state of the tide.

Eilean nan Bairneach and *Eilean nan Trom* from above Roshven Farm

9. **Illan Dou (34). NM 7211 7938.** G. *An t-Eilean Dubh,* 'the black island'. **Nabarnnach; Nabarnock (35).** G. ? *Na Bàirneach* 'the barnacle'. Named *Eilean nam Bàirneach* 'the island of barnacles' on OS maps.

10. **I na Churist (34): Nacharish (35). NM 7323 7928.** The meaning is obscure. Named *Eilean Buidhe* 'the yellow island' on OS maps.

GLENFORSLAN : Gleann Forslan : the glen of the land of the waterfall

1686, Glenforslan (57); 1739, Glenforslan (1); 1744, Glenfordland (38); 1749, Glenforsalan (28); 1792, Glen Forslan (69); 1809, Glenforsland (71); 1845, Glenfersallen (8); 2002, Glenforslan (2).

Valuation: A one merk land in 1749, and let with the half merk land of Duilad at that time (28).

Glenforslan steading, house and waterfall

Tenants: 1686, John McEan VcDugald (57); 1744, Mary Kennedy, daughter of Effie McDonald (38); 1746, Hugh MacVoddich, John Maclean (12); 1748, Margaret Cameron held the merk land of Glenforsalan and ½ merk land of Duilad (28); 1782, Angus MacNeill (65) and John MacLeod (64); 1792, set in tack to Colin Campbell, tacksman of Glenstrae for 15 years (69); 1802, Peter MacIntyre, tenant (70); 1809, Peter Stewart (71).

Settlements: NM 7551 7370. *Glenforslan.* House, barn, byre and tup park.

Shielings: NM 7632 7377. *Glenforslan.* Scattered shieling with a number of buildings in the gravel fan below a ravine, but permanently occupied at times.

Placenames:
1. **Glenforslan (1). NM 7551 7370.** G. Probably *Gleann Forslan* or *Gleann Fhorslan*, with a loan from Old Norse *Forsland* 'the land of the waterfall', with gen. sg. of *fors* m. 'waterfall'. The waterfall is called *Eas Briadha* 'the splendid or showy waterfall', or *An Eas Bàn* 'the white waterfall'.

2. **Duilad (28). NM 7746 7577.** G. Possibly *An Dìollaid* 'the saddle', but this is a cold, north facing township which gets very little sun in winter. Probably derived from *Dubh* 'black', and *leitir* 'hill slope', to give *(An) Dubh-Leitir* 'the black hill slope'. A half merk land let with Glenforslan in 1748/9 (28) and with Assary in 1749.

3. **Beinyahre (66). NNM 7812 7488.** A phonetic rendering of *Beinn Gàire 666m.* 'the mountain of laughter'.

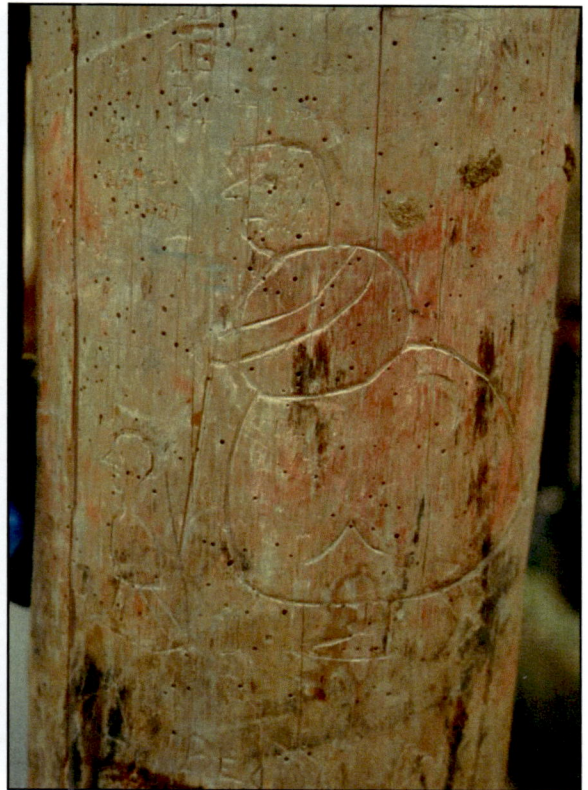

Graffiti in the steading/clipping shed at Glenforslan. The soldier was *Alasdair Glas.*

The shieling site on the gravel fan below the ravine

A house and kail yard at the lower end of the shieling

A line of three buildings aligned against the contour

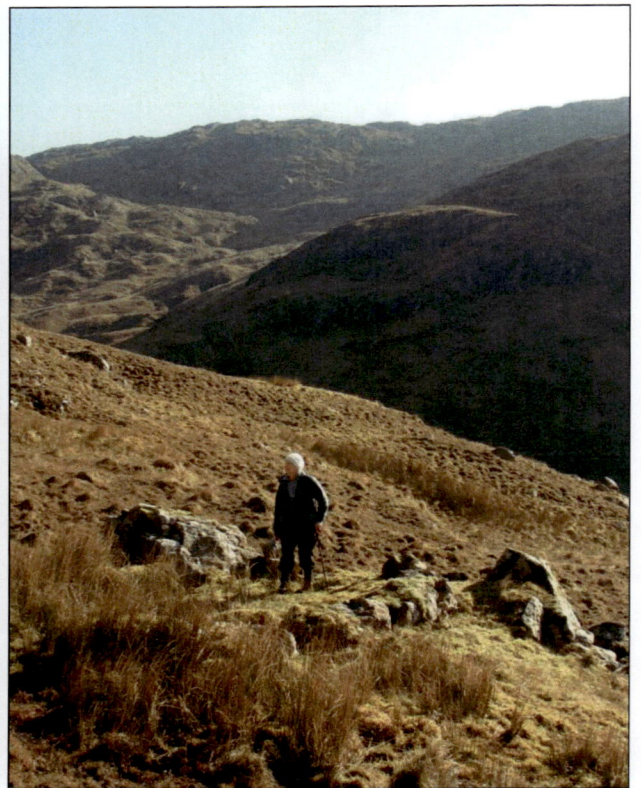

Building stance on a small dry knoll

ASSARY : Àsairigh : the shieling of the ridge (?)

1744, Assary (38); 1748, Afsary (28); 1782, Assary (65); 1786, Assory (68); 1789, Duillad, Assary (63); 1823, Imerassary (7); 1845, Diallod and Assary (8); 2002, Assary (2).

Assary from the east   © Dr. Sandra Evans

Valuation: A ½ merk land in 1748, and a 1 merk land in 1755. Diallod and Assary combined to form a one merk land in 1845.

Tenants: Donald, John and Angus McDonald had an equal share in the ½ merk land of Assary in 1748 (28). 1782, Angus MacLean, John MacPherson, Alexander MacEachern (65); 1786, John MacPherson, John MacEachen, Ewen MacDonald and John Smith (68). 1792, Set in tack to Archibald MacDonald, Duchamis and Donald MacEachern (69). 1802, John MacDonald, Borrodale, tenant (70); 1809, Peter Stewart, Glen Forsland (71).

Settlements: NM 7625 7585. *Assary.* At least 8 buildings noted, but probably many more.
NM 7746 7577. *Diallod ?* At least 8 buildings and a fank.

Shielings: Not recorded.

Placenames:
1. **Assary, (2). NM 7625 7585.** If from G. *Àsairigh* may derive from Old Norse *Àssærgi* 'the shieling of the ridge', with gen. sg. of *áss* m. 'ridge', or perhaps more likely a 'vaccary', 'cattle ranch'. In 1789 it was stated that '*There is not a single stick of wood of any kind upon Duillad or Assary*' (63). The nearest woodland of any value was in Brunary. The Mingary Baptismal Record spells it 'AHASARY'. Professor Watson has the word *Assairidh* 'a rising green slope', and add the word *Ath* 'ford' to give *Ath-Assairidh* 'the ford of the rising green slope'.

2. **Diallod (9); Duilad (28). NM 7746 7577.** G.*(An) Dubh-Leitir* 'the black hill slope' or *(An) Diallaid* '(the) saddle'. *yes*
The location is conjectural, but in 1749 it was a half merkland, let with Glenforslan, the adjoining township. This is a miserable, cold, north-facing township and will see little or no sunshine in winter.

INCHRORY : Innis Ruairidh: Rorie's meadow

1686, Inshrory (57); 1718, Inchraury (6); 1739, Innisruary (1); 1744, Inshrory (38); 1748 Ifiroy (28); 1755, Issiroy (72); 1762, Inchrory (54); 1823, Inchrory (7); 2002, Inchrory (2).

Valuation: A ½m land in 1718 and 1755, a 2 farthing land in 1798.

Tenants: 1686, Finuall nein Dugald VcEan VcRuary (57);1718, Donald McEan vic Coull and Dugall McEan vic Coil (6). 1744, Mary McDonald (38). In 1762, Ewan Macdonald and John Macdonald each held 3 clitichs, while John Corbet Smith had two clitichs = 8 clitichs or a 1 penny land. 1798, Set to Donald Macdonald in Lochans (71).

Settlements: NM 7560 7510. *Inchrorie*. At least 14 buildings, and probably many more.
      NM 7592 7546. Three buildings and a field.

Shielings: NM 7520 7510. Developed shieling within fields, and at least 4 buildings.
      NM 7520 7581.
      NM 7514 7544. Two buildings in a bracken patch.

Placenames:
1. **Innisruary (1). NM 7560 7510.** G. *Innis Ruairidh* 'Ruairidh's meadow', with gen. sg. of the man's name *Ruairidh* 'Rory'. *Innis* 'island; sheltered valley protected by a wood; field to graze cattle in: haugh, riverside meadow'.

SHONA BEAG : Seona Bheag : little Shona :

1749, Shunavegg (28); 1782, Shonabeg (64); 1786, Shonaveg (68); c.1816, Shona Beg (3); 1845, Shewnabigge (9); 2002, Shona Beag (37).

The causeway across to Shona Beag

Valuation: A feorling (¼d) land in 1845.

Tenants: 1748, Angus MacPherson; 1782, Duncan MacEacharn (64); 1784/85, Duncan MacEachen and his brothers (67); 1792, 1802, Archibald MacEachen (69,70).

Settlements: NM 6676 7358. 'Invermoidart'. Scattered group of 8 structures.
      NM 6613 7372. 'Leiter Bhuidhe'.

Shielings: Not recorded.

Placenames:

1. **Shona Beag (2). NM 665 735.** G. *Seona Bheag* 'little Shona', derived from *Sjón* 'watching'' so *Eilean Shona* may be the tautological 'island of Lookout Island'.

2. **Torr an Fiann (4). NM 6614 7363.** A hill name but the specific is obscure. Possibly G. *Tòrr nam Feinne* 'the hill of the giant or warrior' or *Torr nam Fiann* 'the hill of the giants'.

3. **Cnoc an Tarabet (4). NM 6636 7353.** G. *Cnoc an Tairbeirt* 'the hill of the isthmus', with gen. sg. of *tairbeart* m. 'isthmus', or *Cnoc na Tairbeirt* 'the hill of the portage place', with *tairbeart* f.

4. **Chona (4). NM 6675 7335.** G. *Seona* 'Shona'. Named **Invermoidart** on OS maps. See (1) above.

5. **Bealach Chona (4). NM 6676 7359.** G. *Bealach Sheona* 'the pass of *Seona* (lookout island)'.

6. **Caolas Chona Bheag (4). NM 6687 7400.** G. *Caolas Sheona Bheag* 'the straits of *Seona Bheag*'. Also known as *Caolas Iain Oig* and *Cùl Eilean Seona*.

7. **Rudha Coire Sneaghd (4). NM 6635 7398.** G. *Rubha Coire an t-Sneachda* 'the point of *Coire an t-Sneachda* (the corry of the snow)', with gen. sg. of *sneachd* m. 'snow'. *Coire an t-Sneachta* is an odd name as there is hardly a *coire* here, it is unlikely to hold much snow and the hill above only rises to 100m.

8. **Meall Coire Sneaghd (4). NM 6635 7398.** G. *Meall Coire an t-Sneachda* 'the hill of *Coire an t-Sneachda* (the corrie of the snow)', with gen. sg. of *sneachd* m. 'snow'.

9. **Eilean Gruachan (3). Eilean Gruachach (4). NM 6650 7306.** G. *Eilean a' G ...?*, 'the island of the ...?' The name appears to have been entered in pencil on the estate plan, then inked in at a later date. The meaning of the word is obscure. Possibly G *Gruaigean*, the seaweed bladderwrack or henware is intended here.

10. **Illan an Ach (3); Eilean na Fiach (4). NM 6710 7285.** G. *Eilean an Ach* 'the island of the meadow', with a shortened form of the gen. sg. of *achadh* m. 'meadow'. Possibly *Eilean na Fitheach* 'the island of the raven'. Named *Eilean an Fhèidh* 'the island of the deer (sg)', with gen. sg. of *fiadh* m. 'deer' on OS maps.

11. **Sgeir Nighean Sheumais ((23). NM 6694 7296.** G. *Sgeir Nighean Sheumais* 'the skerry of *Seumas*'s daughter', with gen. sg. of the man's name, *Seumas* , 'James'. When John of Moidart was Captain of Clanranald, about 1584, two men stole some silver coins from the castle, and after a brief trial, were found guilty and executed on Gallow's Hill, *Cnoc a' Chrochaidh*. The coins were never recovered, but James's daughter was thought to know where they had been hidden. As she refused to reveal their whereabouts, she was tied to this skerry until drowned by the rising tide.

12. **Shunavegg Wood (28).** The woodland was valued at one pound in 1748/9. Much of the island is under tree cover today and it is impossible to place the woodland at that time.

13. **Airthrago (19).** An early name for *Eilean Seona* which appeared in Adamnan's Life of St. Columba and discussed in Watson (1926). The name is derived from G. *Air Thraigh* 'on shore', but the archaic meaning of *Air* was 'before', hence 'before shore' due to the island being tidal.

Shona Beag and *Eilean an Fheidh* from Briaig   © Dr. Sandra Evans

LANDS IN MOIDART BELONGING TO MACDONALD OF GLENALLADALE, 1739.

GLENALADALE: Ölrdal : Alder Valley

1617, Glenarradill; 1674, Glenalladell (11); 1686, Glanalvadill (57); 1748, Glenalladel (32); 1823, Glenaladale (7); 1832, Nalla dale, (34); 2002, Glenaladale (37).

Glenaladale from Gorstanvorran

Settlements: NM 8212 7724
NM 8226 7615.
NM 8235 7527. Fank.
NM 8243 7513.

Valuation: A two merks ten shilling land in 1674 and a 2 merks and Ten penny land in 1748 (32).

Tenants: 1674, wadset (?) to Rorie McDonald of Glenalladale (11); 1744, Ewen McVarish and Beag McGilvra his spouse (38); 1746, Sandie and Allan McDonald, Angus Mor McDonald, ? MacIan vic Ian, Rorie MacDonald, John MacDonald, Angus MacVorrich (12); 1748, Alex. McDonald of Glenalladel Feuar (32); 1782, John Smith (64).

1. **Glenaladale (2). NM 8246 7514.** G. *Gleann A...,* 'the valley of A...', with a loan ultimately from Old Norse –dalr 'valley'. As the pronunciation of the Gaelic is uncertain, it is difficult to be more specific, but ON *Ölrdal* 'Alder Valley' is a possibility. The haughland near the loch is still covered in old Alder trees, a good example of Alder Carr, and an important component of the Loch Shiel SSSI.

On 10th June, 1617, a complaint was made to the Privy Council by *Donald McAllan VcEane of Ylantyrum, Captain of the Clanronnald* against a number of Clan Cameron men who had raided in Moidart during the previous October, and ill treated a number of the inhabitants:-
On ...October last, Ewne McEane Twiche in Tyrlandie (Torlundy), Allane McCharlis McEwne McEane in Mamor (Mamore), and Lauchlane McChairlis VcEwne, his brother, with others to the number of twenty four, all armed, went to (the) persuer's lands of Glenarradill in Moidert, and not only took away the hail nolt (cattle), horse, shep and gait (goats) and uther bastial, to the number of 400 animals in all, belonging to his tenants, but also attacked Katherine Nine Eane Duy (daughter of Iain Dubh), spouce of Johnne McAngus McCoule in Glenarradill, Everick Nin Gillichallum, spouce of Dougall McAngus Moir there, struck them on sindry parts of their bodies, bound them with 'widdeis and bowstringis, and carried them away with sixteen of the persuer's tenants, men and women. They conveyed them the distance of 16 miles, having bound them all with bowstrings and tirred (stripped) them naiiked of all their clothes, exposit thame to the injure of the cauld weddir, qhairby Moir Nin Elspik Roy, on the way as sho wes caryed bound away with thame as said is, partit with bairne, to

226

the extreame hasard of Hir lyff. Persuer appearing by James Logy, and defenders not appearing, the Lords order said defenders to be pronounced rebels. (Register of the Privy Council, First Series, Vol. XI, P. 149).

Glenaladale appears to have been divided into several townships or settlements that were let out to various tenants. There is very little documentation for these, and their names and marches are conjectural. They are, or may have been,

BRYNAFERN; DELROYD; SALNOES, COIRE NAM MAIGHEACH and ALLT AN DOBHRAIN.

BRYNAFERN: Bràigh na Feàrna : the brae of the alder
1739, Brynafern (1); 1748, Drynafern (31); 1823, Glenaladale (7), 1832, Derrynafern (34).

Settlements: NM 8162 7814. Fank.
NM 8170 7810. Fank.
NM 8180 7810.
NM 8216 7724. Three buildings.
NM 8179 7786. One building recorded.

[handwritten margin note: see Dail nar Brae / Bràigh na fearn. / local dialect use sing. for plural]

Shielings: Not recorded.

Placenames:
1. **Brynafern (1). NM 8181 7802.** G. *Bràigh na Feàrna* 'the brae of the alders', with gen. sg. of *feàrn* f. 'alder'. It is difficult to relate Roy's Map to the OS maps, so the next two names are 'best guesses'.

2. **Ben Sneacht (31). NM 8282 7880 ?** G. *Beinn an t-Sneachda* 'the mountain of the snow', with gen. sg. of *sneachd* m. 'snow'. Named *Beinn Mhic Cèdidh* and *Beinn na Cuaiche* 'the cupped hill' on OS maps.

3. **Ben mich adch (34). NM 8282 7880.** G. *Beinn 'ic Cheididh..* Named *Beinn na Gaoithe* on the 1st Edition OS map.

4. **Ben Oir (31). NM 8072 7946 ?** G. *Beinn an Oire* 'the ben of the boundary' or *Beinn an Òir* 'the mountain of the gold', with gen. sg. of *òr* 'gold'. Named *Diollaid Mhòr* 'the great saddle' or 'the great black hill slope' on OS maps.

5. **Scur an Hurranad (34). NM 7950 7780.** The ridge on the south side of *Coire Reidh.*

DELROYD : Dail na Roid: the meadow of the bog myrtle
1739, Delroad (1); 1748, Dalroyd (31); 1832, Dalrayd (34).

The western end of Dail na Roid

Settlements: NM 8010 7265.   Dalroyd. Three structures recorded.
NM 8137 7335. 'Gasgan Wood'. One building recorded.
NM 8236 7498. One building recorded in a small field system.

Shielings: Not recorded.

Placenames:
1. **Delroad, (1). NM 820 738.** G. *Dail na Roid* 'the meadow of the bog myrtle', with gen. sg. of *roid* f. 'bog-myrtle'.

Dalroyd from Rubha an Fhaing Dhuibh

2. **Ben Gaar (31). NM 81 74.** An obscure name although it might be for Gaelic *Beinn Ghobhar* 'goat's mountain', with gen. pl. of *gabhar* m. 'goat' or *A' Bheinn Gheàrr* 'the short mountain', but cf. *Beinn Gàire* to the west, literally 'laughter mountain', perhaps from an echo (NM 781 749). Named *Beinn an t-Samhainn* 'the mountain of Hallow-tide' or 'November' on OS maps. The hill retained good grass until late in the season and the name implies 'late grazing'.

3. **Cruach na Fnarach (34). NM 8124 7449.**

4. **Croch na curiagh (34). NM 7920 7375.** Possibly just a phonetic rendering of *Cruach nan Caorach* as on the OS Maps.

5. **Rubha nam Braithre (19). NM 8188 7361.** G. *Rubha nam Braithre* 'the point of the brothers'.

6. **Rubha nam Peathraichean (19). NM 8205 7374.** G. *Rubha nam Peathraichean* 'the point of the sisters'.

Dail Roid, beyond the trees in the middle distance

228

GASKAN : Gasgan : fish-tail shaped estuary (?)
1739, Gaskan (1); 2002, Gaskan (2).

Gaskan House at the foot of the Allt Ghuibhais, the march between Annat and Dalroyd

Settlements: NM 8011 7267, *Gasgan.*

Shielings: Not recorded.

Placenames:
1.   **Gaskan (1). NM 8011 7267** G. *Gasgan* 'fish-tail shaped estuary', or possibly 'plateau', 'copse' or 'the tail' (shaped portion of land). Gaskan may have been a settlement within Dalroyd in 1739, but the present Gasgan forms part of Annat. There is a half remembered oral tradition in the district that Clanranald's birlinns were built here, then taken to a *port* on Briaig or Port a' Bhata for fitting out.

SALNOES : ? Sàil an Òis : the point of the river-mouth
1748, Salnoes (31); 1832, Saluclis (34).

Glen Aladale from Torr a' Choit

Settlements: NM 8243 7507.
                   NM 8233 7484.
                   NM 7237 7491. One building .

Shielings :   Not recorded.

Place-names :

1. **Salnoes (31). NM 8242 7430.** This might possibly represent G. *Sàil an Òis* 'the promontory of the outlet', with *sàil* f. 'heel, point' and gen. sg. of *òs* m. 'river-mouth, outlet'. *Sàil an Ois* 'the promontory of the stag' and *Sàil an Nis* 'heel of the promontory' have also been suggested.

2. **Ben na cairn (34). NM 8395 7724.** *Beinn na Carn* 'the mountain of the cairn'. Named *Beinn a' Chaorain* 'the sheep mountain' on OS maps.

3. **Ben niy (34). NM 8348 7601.** Named *Beinn Bhuidhe* 'the yellow mountain' on OS maps, and Ben niy is probably a phonetic rendering of this.

4. **Innis Ard nan Clach (60). NM 8553 7675.** G. *Innis Ard nan Clach* 'the meadow of the height of the stones'. *Innis* 'a sheltered valley protected by a wood, favourable pasture, a riverside meadow'.

5. **Allt Geal (60). NM 8574 7700.** G. *Allt Geal* 'the white burn'.

6. **Rubha na Càraid (73). NM 8474 7599.** G. *Rubha na Càraid* 'the promontory of the couple' or 'the lover's point'. *Caraid* 'male friend or relation; cousin' or *càraid* 'pair, couple, twins, married couple'. A couple were drowned here in the early 19th century, and the name commemorates the event.

7. **Rubha Sgairt a' Choilich (19). NM 8525 7653.** G. *Rubha Sgairt a' Choilich* 'the headland of the cleft of the black-cock'.

8. **Rubha an Uillt Ghil (19). NM 8587 7688.** G. *Rubha an Uillt Ghil* 'the headland of the white burn'.

9. **Rubha nam Fiadh (19). NM 8602 7703.** G. *Rubha nam Fiadh* 'the headland of the deer'.

Beinn Odhar Bheag on the march with Creag Bhrodainn

Beinn a' Chaorainn with Sgèir Dhuibh Camas an Eireannaich (left)

ALLT AN DORAIN : Allt an Dòbhrain : the burn of the otter

1739, Aultndorain (1).

Tenants: 1746, Duncan MacInnish vic Ian (12). *Duncan son of Angus son of John.*

Settlements: NM 8175 7597. *Allt an Dòbhrain.* Fank.
               NM 8186 7646.
               NM 8176 7605. One building recorded.
               NM 8175 7642. 'Allt a' Choire Mhòire'. Four buildings recorded.

Shielings: Not recorded.

Placenames:

1. **Aultndorainn (1). NM 8175 7597.** G. *Allt an Dòbhrain* 'the burn of the otter', with gen. sg. of *dòbhran* m. 'otter'.

2. **Bad an Dobhrain (1). NM 8175 7597.** G. *Bad an Dòbhrain* 'the thicket or place of the otter', with gen. sg. of *dòbhran* m. 'otter'.

CORRYNAMAICH : (?) Coire nam Maigheach : the corry of the hares.

1748, Corrynamaich (31).

Roy's Military Survey of 1748 is the only source for this name. It may have been a settlement within the township of *Allt an Dobhrainn.*

Settlements : NM 8184 764.

Shielings : Not recorded.

Place-names :

1. **Currynamaich (31). NM 8184 7645.** Possibly G. *Coire nam Maigheach,* 'the corry of the hares', or *Coire nan Each* 'the corry of the horses', with gen. pl. of *each* m. 'horse'.

2. **Màm Lorgan (73). NM 8117 7607.** G. *Màm na Luirg* 'the hill path'.

3. **Dail nam Maigheach (73). NM 818 760.** G. *Dail nam Maigheach /Margheach* ' the haugh of the hares'.

231

CREAG VRODDIN : Creag a' Bhrodainn : the rock of the goats (?)

1694, Craigvroddin (58); 1739, Creignrodinn (1); 1744, Craigvroddin (38).

Beinn Odhar Bheag, Sgurr an Tuirc and Beinn Odhar Mhòr

Tenants: 1694, Agnes, spouse to Allan McDonald, with Mr Alexander, Sorle, Ronald, Rorie, Angus, Archibald, James and Janet, children (58); 1744, John Smith and Mary McDiarmid his spouse (38).

Settlements: Several small settlements just up from the shore but none appear in the archaeological record.

Shielings: Not recorded but much of the hillside is very steep and rough, and some of the later settlements may have developed out of shielings.

Beinn Odhar Mhòr from Glaschoirean looking north across Loch Shiel

Placenames:

1. **Creigvrodinn (1).** Perhaps for Gaelic *Creag a' Bhrodainn* 'the rock of the goad or staff'. About 300m above is *Meall a' Bhrodainn* with a rock pinnacle on top, reminiscent of a goad. Gordon Barr suggests *Creag a' Bhrodainn* 'the crag of the goats', with *brodann* Gen, *brodainn* 'goat', which seems more likely. *Go and see the pinnacles.*

2. **Torr a' Phrionnsa (73). NM 865 776.** G. *Tòrr à Phrionnsa* 'the hill of the Prince', with gen. sg. of *prionnsa* m. 'prince', where Bonnie Prince Charlie stopped on his way to Glen Finnan in August, 1745.

3. **Scor Vrottan (31). NM 85 79.** Perhaps G. *Sgor a' Bhrodainn* 'the cleft of the goad'; cf. no. 1, above. Named *Beinn Odhar Mhòr* 'the large oatmeal-coloured mountain' on OS maps.

4. **Emdluigh (34). N 8700 7837.** The Gaelic is obscure. Possibly named *Maol a' Bhrodainn* on OS maps.

5. Allt nam Breac (60). **NM 8600 7736.** G. *Allt nam Breac* 'the burn of the trout'. *Breac* 'trout'. Named *Allt na Brachd* 'the burn of the putrefaction' on OS maps. *Bràchd* fem. sub. 'putrefaction, fermentation'.

6. **Càrn MhicCuaraig (73). NM 8774 7873.** G. *Càrn MhicCuaraig* 'Kennedy's cairn'.

7. **Creag Anna (73). NM 8737 7896** G. *Creag Anna* 'Anna's crag'.

8. **Clach a' Phuist (19). NM 8622 7803.** G. *Clach a' Phuist* 'the stone of the post'. An overhanging rock where the postman left the mail for the Glenaladale people to collect. *shepherds*

9. **Rubha Allt na Bracha (19). NM 8627 7723.** G. *Rubha Allt na Bracha* 'the burn of the flash flood'. Possibly 'the burn of the putrefaction'. *Allt na Brachd on OS. Pathfinder series. Another form of Bruchd – flash flood. Applies*

Coille Bhrodainn

233

GLEN FINNAN : Gleann Fionn Abhain : Glen of the White River
1674, Glenfinen (11); 1739, Gleninnin (1); 1748, Glen Invin (32); 1823, Easter and Wester Finnan (7); 2002, Glenfinnan (2).

Dail an Tùir

Glen Finnan from Rubha nan Gamhna

Valuation: A thirty shilling land in 1674 and in 1748.

Tenants: 1674, wadset (?) to Rorie McDonald of Glenalladale (11). 1745, Alexander MacDonald, Ranald his brother, Allan Du, Allan Mac Allister vic Murdo, Ewan Mac Millan, John, his brother, Angus More, Angus MacPherson, tailor and Angus MacIsaak (22); 1746, Angus Mor, Angus MacPherson, Ewan MacMyllan and his brother John, Angus MacIsaac, Ranald MacDonald (12); 1748, Alex. McDonald of Glenalladel Feuar (32).

Settlements: NM 900 806. Slatach.

Shielings: Not recorded.

Placenames:

1. **Glen Finnan (2). NM 900 806.** G. *Gleann Fionn Abhain* 'the glen of the white river' or *Gleann Fhionghain* '*Fionghan*'s Glen', with gen. sg. of the man's name *Fionghan* 'Fingon'. Possibly G. *Glean Fhionnainn* '(St) Finnan's Glen'. *Named after the river, frequently in spate; is white water nearby all the way*

2. **Allt an t-Sluicht (19, 34). NM 8800 8010.** G. *Allt an t-Sluichd* 'the burn of the cleft', with gen. sg. of *slochd, sloc* m. 'gully, hollow, pit, cleft'.

3. **Loch an t-Sluichd (19, 34). NM 874 798.** G. *Loch an t-Sluichd* 'the loch of the hollow'. Named *Lochan nan Sleubhaich* 'the lochan of the wilderness' on OS maps.

4. **Water of Finnan (13). NM 9095 8120.** G. *Fionn Abhainn* 'the white river'. River Finnan on OS maps.

5. **Allt na nua (13). NM 8990 8067.** G. *Allt nan Uamh* 'the burn of the caves', but if the land nearby was good grazing, then *Allt an Utha* 'the burn of the udder' is more likely. *The hill of the same named looks like an udder upside down.*

6. **Camus Aylihou (13). NM 8817 7865.** G. *Camas A...?* 'the bay' with an unknown specific. *Còimhlaidhean Bàthur Bay. See Kinlochm*

The heather-capped Eilean Dubh

Choimhleachan – Hugh MacLeod

7. **Eilean Dou Camus Aylihou (13). NM 8817 7856.** G. *Eilean Dubh Camas A...?* 'the black island of Camas A'. The Gaelic is obscure; could Aylihou be an anglicisation of *Eilean Dubh*, with the name translating as 'the black island of the bay of the black island?'.

8. **Schardben (34). NM 9468 8635.** Possibly an alternative name for *Streap* on OS maps. See No. 27.

9. **Bangery (34). NM 9028 8800.** Possibly an alternative name for *Sgùrr nan Coireachan* as on OS maps, although it sounds like a shieling/vaccary name, with *bàn* 'white, fair' and *airidh/ærgi* 'shieling, vaacary'.

10. **Brain Kyle (34), Breun-choille (19). NM 8843 8175.** G. *Breun-choille* 'the filthy wood'. This was a farm or croft c. 1790.

11. **Doielach (34).** G. *An Diollaid* 'the black slope' or 'the saddle'.

12. **Port a' Bhàta Mhòir (73). NM 8903 7944.** G. *Port a' Bhàta Mhòir* 'the port of the big boat'.

13. **An t-Eas Smúid (73); Eas na Smùid (19). NM 916 809.** G. *An t-Eas Smúid* 'the smoky falls', 'the falls of the vapour'. There is a lot of spray when the river is in spate.

14. **An Gabhal Mòr (73, 19). NM 905 810.** G. *An Gabhal Mòr* 'the big fork or enclosure'.

Slatach from the head of Loch Shiel

River of poles.
Abhainn Shlataich

15. **Taynaslatich (31). NM 8991 8045.** G. *Taigh na Slataich* 'the house of *An t-Slatach*', with gen. sg. of *slatach* f. 'reedy place' a 'willowy place' or 'house of the rod-place'. Could the name infer a creel house? Prince Charles Edward Stuart lodged in an earlier house here during the period he was in Glen Finnan in August 1745. The name may be derived from the nearby *Abhainn Shlataich* 'River of Poles'. *Yes.*

16. **Stronnaguine (34). NM 8918 8020.** G. *Sròn na Guine* 'the point (nose) of the woman'. Probably an alternative name for *Druim na h-Aire* as on OS maps. *Sròn an Ath Dhuinn? Brown Ford Point – describes the river mouth.*

17. **Ault na mia (34). NM 8700 8147.** A name given to the middle section of the *Abhain Shlatach.*

18. **Leac na h-Uinneige (73). NM 912 843.** G. *Leac na h-Uinneige* 'the slab of the window'.

19. **Eilean na h-Uinneig (19). NM 9175 8440.** G. *Eilean na h-Uinneig* 'the island of the window'.

20. **An Uinneag (19). NM 9118 8425.** G. *An Uinneag* 'the window'. Nos. 19, 20 and 21 are names derived from a hole in a rock worn by the river.

21. **Dail an Tùir (73. NM 9065 8062.** G. *Dail an Tùir* 'the haugh of the tower' i.e. monument.

22. **An Teanga Mheadhoin (19). NM 898 818.** G. *An Teanga Mheadhoin* 'the middle tongue'.

23. **Camas Dubh nam Bratach (19). NM 8795 7860.** G. *Camas Dubh nam Bratach* 'the black bay of the banners'.

24. **Camas Choimhlachain (19). NM 8817 7865.** G. *Camas Choimhlachain* 'the bay of the little barrier'. The bay is obstructed by an island and gravel spit which is almost dry when the water is low.

25. **Tom an t-Saighdeir (19). NM 9163 8345.** G. *Tom an t-Saighdeir* 'the soldier's knoll'.

26. **Sron a' Chuirn (19). NM 9105 8219.** G. *Sròn a' Chuirn* 'the nose of the rocky place'.

27. **Na Righeannanaich (19). NM 9300 8600.** G. *Na Righeannanaich* 'the shielings'. *The mountain base.*

29. **An Garbh Buaile (73, 19). NM 8103 8125.** G. *An Garbh Buaile* 'the rough fold'. This was a holding park for sheep and cattle

30. **Eilean Dharaich (19). NM 9126 8169.** G. *Eilean Dharaich* 'oak-tree island'.

31. **Tom na Drochaid (19). NM 9178 8466.** G. *Tom na Drochaid* 'the hillock of the bridge'. *(bridge shape)*

32. **Cnap na Laoigh (19). NM 9230 8319.** G. *Cnap nan Laoigh* 'the knoll of the calves'.

Dail an Naoimh: the burial ground   *Dail Naoimhe - Hallowed Field.*

33. **Dail an Naoimh (73); Dail Naoimhe (19). NM 9060 8064.** G. *Dail an Naoimh* 'the field of the saint' or 'the hallowed field'. An old burial ground close to the possible site of the 'Raising of the Standard'. See below. Was this site chosen by the MacDonalds so as to add sanctity to the Prince's cause? *NO. Graves are older.*

The Higher Pass.

34. **Bealach nas Airde (19). NM 8850 8525.** G. *Bealach nas Àirde* 'the pass of the ?'. **Beloch na Shard** on Roy's map.

35. **Camas nam Bradan (19). NM 893 798.** G. *Camas nam Bradan* 'the bay of the salmon'. *also Camas an Thaing.*

36. **Leum an t-Saighdeir (19). NM 9121 8318.** G. *Leum an t-Saighdeir* 'the soldier's leap'.

The head of Loch Shiel (Detail).   **'Here Prince Charles's Standard was Placed'.**
The site of the Raising of the Standard at Glen Finnan on 19th August, 1745, according to William Bald.

National Records of Scotland, RHP72/1-8, Plan of Ardnamurchan and Sunart, Argyll, 1806.

237

LANDS IN MOIDART WADSET TO MACDONALD OF DALELIA, 1739.

Leith Half. – divided by ditch – (handwritten)

DALELIA : Dail an Léigh : the field of the physician

1625, Derrilea (30); 1686, Daliell (57); 1707, Daleilve (9); 1718, Dellila (6); 1739, Dallala (1); 1748, Dalellie, Dallellie (28); 1790, Daleley (48); 1823, Dalelea (7); 2002, Dalelia (37).

Dalelia House

Valuation: A 20/- land (=1½ merkland) in 1625 and a 1½ merk land in 1718, 1748 and in 1790.

Area: C. 1816 there was 59.05 acres of arable, 11.27 of plantations and 1239.37 of moorland, total 1309.69 acres.

Tenants: 1625, John ronnaldsoun, persoun of Ellanfinnan (30); 1686, Effrick nein Donald VcEan VcWilliam, spouse to John McEan VcRuary; Anna nein Neill VcEan, spouse to John McEan VcRuary (57); Wadset to Maighstir Alasdair McDonald who died in 1718, then to his son Angus Beg (Aneas) McDonald of Dallellie. 1744, John McInnins Ferryer at Ellenfinand (38). Held in tack by Alexander Macdonald of Dalelia for 31 years from November 1790 (48).1813, sold to Alexander Macdonald of Glenalladale (49).

Settlements: NM 7842 6933. *Dalelia.* House and kailyard in 1816.
NM 7443 6885. *'Ru Ard Core'.* One structure and a small field.
NM 7233 6968. *Austin's Croft or Croit Uisdein.* House and kailyard.
NM 7504 6845. *Port an Eilean.* House, byre and small field. Probable motte.
NM 7209 6961. *'Rubha na Croite'.* Two structures but no fields.

Shielings: Not recorded.

Placenames:

1. **Dalelia (2). NM 7342 6933.** G. *Dail an Léigh* 'the field of the doctor' , with gen. sg. of *léigh* m. 'doctor'. Iain MacMaster, Mingary has suggested *léigh* 'a Master of Arts', while Ronald Black suggests *Dail Eilghidh* 'a field levelled for ploughing'. Local tradition suggests G. *Dail Leithe* 'the half meadow' or 'the halved haugh', while the 1625 spelling 'Derrilea' suggests G. *Doire Liath* 'the grey grove'.

2. **R. McKeman (3). NM 7207 6925.** G. *Rubha MhicEamoin* 'Eamon's son's headland' or *Rubha MhicÀmainn* '*MacÀmainn*'s point', with gen. of the personal name, although this assumes it had a Gaelic form. McKeman is an Anglo-Irish form and if the owner of the name was not from the area, particularly if he was not a Gaelic speaker, the English form would likely have been used. (Cf. Black, *The Surnames of Scotland.*, MacCam(m)on(d); Kneen, *Manx Personal Names,* MacKemayn, Keman, Keymaine, Kemaine*)*. Named *Rubha na Croite* 'the headland of the croft' on OS maps (2).

MacEamoin – descended from Kathleen O'Briain (handwritten)

3. **Ru Man na Bann (3). NM 7241 6916.** The form Man here is likely to be a copying error for Illan and we should therefore read G. *Rubha Eilean nam Ban* 'the point of *Eilean nam Ban*'. Named *Rubha nan Coinnlean* 'the point of the barley stalks' on OS map (2).

4. **P. Dalelia (3). NM 729 6909.** G. *Port Dail an Léigh* 'the landing place of *Dail an Léigh*', see no. 1, above.

Rubha na Fainge

5. **Ru Fang (3). NM 7357 6886.** G. *Rubha na Fainge* 'the point of the fank', with gen. sg. of *faing* f. 'fank, sheepfold'.

6. **Ru Ard na Monagh (3). NM 7390 6854.** G. *Rubha Àird na Mònadh* 'the point of the headland of the peat', with gen. sg. of *mòine* f. 'peat'. Named **Ru Ard na Mollagh** on Ms 384 (1) and *Rubha na h-Àird Mholaich* 'the point of the rough (or wooded) headland' on OS maps (2).

7. **Sadil na Cull (3). NM 7442 6860.** G.? *Sàideal a' Chaoil* 'Cuill bay' or 'the bay of the narrows'.

8. **Ru Ard Core (3). NM 7446 6846.** G. ?*Rubha Àird nan Corra* 'the point of the herons headland'. *Àird* 'high' is not appropriate here as the headland is low lying. Named *Rubha nan Corra* 'the headland of the herons' on OS maps. In the early 1800s there was a croft here, with 1.75 acres of arable land.

9. **Ru A Uock karch (3). NM 7527 6851.** G. Probably short for G. *Rubha na h-Àirde Uachdaraich* 'the point of the upper headland', with gen. sg. of *An Àird Uachdarach* 'the upper headland' or it might imply that it is higher than its neighbours.

10. **Craig an Isog (3). NM 7544 6918.** G. ?*Creag an Uiseag* 'the rock of the lark'. Named *Creag an Uisge* 'the rock of the water' on OS maps.

11. **Cruach na Tolach (3, 34). NM 7468 7146.** G. *Cruach na Tollach* 'the prominent hill full of holes'. This appears to be on the Dalelia/Drumloy/Lochans march.

12. **Ru Illan na Bann (3). NM 7240 6917.** G. *Rubha Eilean nam Ban* 'the point of the island of the women', with gen. pl. of *bean* f, 'woman'.

13. **Croit Uisdean (24). NM 7233 6968.** G. *Croit Ùisdein* 'Ùisdean's croft'; 'Hugh's croft', with gen. sg. of the man's name Ùisdean 'Hugh' (or, as suggested by the map form, 'Austin', and named **Austin's Croft** on OS maps (2). Father Ùisdean MacDonald was the Roman Catholic Priest of Moidart 1769-1787. He entered the Scot's College in Rome in 1757, aged 13, and was entered in the college register as Austen (=Augustine), the nearest English sounding name to the Gaelic *Ùisdean,* a synonym of *Eoghann* or the English 'Hugh' in Argyll (59).

14. **Toll Odhar. NM 730 703.** G. *An Toll Odhar* 'the dun-coloured hole'.

Port an Eilean croft, noost/boat shed, jetty and probable motte

15. **Port an Eilean (5); Port an Illen (3). NM 7504 6845.** G. *Port an Eilein* 'the landing place of the island', with gen. sg. of *eilean* m. 'island'. A croft traditionally occupied by the Green Isle ferryman, with one field containing 1.6 acres of arable land.

Post-mediaeval burial grounds for Moidart (left) and Sunart (right) on 'The Green Isle'.

16. **Green Isle (15). NM 7525 6825.** G. *Eilean Fhìonain* 'St. Finan's Isle' or literally, *Eilean Uaine.* The Episcopalian priest, Maighstir Alasdair (MacDonald) lived with the blacksmith here c. 1688, then in a house built by his parishioners, although he stated that he never made use of the land, having no stock at that time. The island was computed at 4.22 acres of quite sweet ground in the early 1800s, but now there are burials throughout, and terraces of the raised beach on all sides, which were used for iron and bronze smelting, stances for creel houses, over 30 recessed platforms for round houses in the mediaeval period (?) and a possible holy well or spring.

17. Allt corrie na clach (34). NM 7377 7000 (?) *Allt Coire na Clach* 'the burn of the corrie of the stone'. It is unclear as to which burn this refers, and '*Coire na Clach'* has not been located.

18. **Rathad Rinean (23). NM 7325 7100. ?** This is the name given to the old road from Kinlochmoidart to Dalelia. 'Rinean' appears to have been a croft name, but as it contains the generic term *Ruidhe,* it may well be a developed shieling.

19. **Arddalie (30). NM 7454 6858** or **NM 7506 6945 (?)** G. *Àird Dail an Lèigh* 'the headland of the meadow of the doctor', with gen. sg. of *lèigh* m. 'doctor'. A 1 merk land in 1625, possibly Cuil and/or Port an Eilean on the eastern side of Dalelia, although it may not be part of Dalelia.

20. **Badenraash (77). NM 7225 7017.** G. *Bad an Rais* 'the place of the scrub'. The estate keeper lived here at one time and the site is also known as '**The Kennels**'

Early mediaeval (?) crosses on *Cnoc an Aingil, Eilean Fhìonain*

21. **Cnoc an h-Aingeal (9). NM 7521 6834.** G. *Cnoc an Aingil* 'the hill of the angel' or 'fire', with gen. sg. of *aingeal* m. 'angel' or 'fire'. A low mound on *Eilean Fhìonain,* where the early (?) crosses were erected. Charles MacDonald (22) records these as having been hewn by Donald Mòr MacVarish of Mingary, who died c. 1840, but similar crosses in Ireland have been attributed to the 10th century or earlier. The tall, slender, fine-grained sandstone cross appears to have come from Carsaig on Mull or Innemore, Morvern, and must surely be set into a socket stone, hardly the work of Donald Mòr. *Cnoc an Aingil* is a frequent place-name in old burial grounds and if 'hill of the fire', may reflect a pre-Christian (?) beltane fire site or beacon point. The presence of the crosses and mediaeval grave-stones within a small enclosure suggests that the rectangular foundation here may have been an early oratory, making this a particularly venerated site and a prestigious place in which to be buried.

Field system above the settlement

Loch Shiel from the east

Possibly the site of 'Rinean'.

22. Clach an Aoig (78). NM G. *Clach an Aoig* 'the skeleton stone'. A late 17th or early 18th C. grave-slab on the south side of *Eilean Fhìonain*, and said to cover the mortal remains of *Maighstair Alasdair,* the Rev. Alexander MacDonald, Episcopalian Minister of Islandfinan, who died shortly before August 9th, 1718 (61). The inscription is very weathered, but has been identified by the Rev. Mike Laybourne, Strontian as from the Book of Revelation, Ch. 14 v.13. No names or dates have been inscribed, but the initials at the top, D MD suggest that the stone was made for a Donald MacDonald, possibly of the Kinlochmoidart family, although the biblical quotation does seem appropriate for an Episcopalian minister.

The Dalelia Mill

Crucifix/reliquary on *Eilean Fhionain*

LANGAL : Lang-vǫll : the long field

1625, Auchnellan ? (30); 1686, Langoll (57); 1718, Langwell (6, 61); 1739, Langall (1); 1748, Langoll (32); 1782, Langale (65); 1790, Langaal (48); 1813, Langall (49); 1823, Langal (7); 1832, Langwall (63); 1851, Langall (9); 2002, Langal (2).

Langal from the NE   © Dr. Sandra Evans

Valuation: A 1 merk land in 1718, 1748 and 1790.

Area: C. 1816 there was 33.15 acres of arable in 8 fields, 161.82 of moss and 718.88 of hill pasture, total 913.88 acres.

Tenants: 1625, John Ronnaldsoun, persoun of Ellanfinnan (30); 1686, Katherine nein Phadrick (57); Wadset to the late Maighstir Alasdair McDonald, who died in 1718, then to his son Aneas McDonald of Dallellie (*Aonghas Beag*). 1746, Allan McDonald, Donald MacEachen (12); 1782, Ronald MacDonald (65); 1782, Ranald McDonald of Daliburgh . Held in tack by Alexander Macdonald of Dalelia in 1790 (48). Sold to Alexander Macdonald of Glenalladale in 1813 (49).

Settlements: NM 7100 6982. *Langal, '* Long Field'.
                NM 7025 6959. *Dail nam Breac,* 'Field of the Trout'.
                NM 6930 7270. 'Auchnellan'   *Meall an Aoil* (?)

Shielings: NM 6936 7221.
                NM 6974 7106. '*Glac Dhubh*'. Developed shieling with a small field system.

Placenames:

1. **Langal (2). NM 7100 6982.** G. *Langal,* from Old Norse, *Lang-vǫll* acc. 'the long field' or 'the long plain'.

2. **Ru Chuin (13). NM 7062 6894.** G. *An Rubha Cumhann* 'the narrow promontory', as on OS maps, but the topography is wrong for such a name.

3. **Ru Gorine (13). NM 6984 6831.** G. *Rubha G....?,* 'the promontory of *G',* with an unknown specific..

4. **Mineral Well (13); Mineruliver (13). NM 7030 6900.** The shore is stained red with the deposition of iron salts for a distance of 100m or more, and no doubt a ready source of bog iron in the past. A chalybeate Spring.

5. **Grianan (24).** NM 700 696. G. *An Grianan* 'the sunny place' or 'the hill'.

6. **Innis a' Rubha (22).** NM 6980 7214. G. *Innis an Rubha* 'the meadow of the promontory', with gen. sg. of *rubha* m. 'promontory'.

7. **Maul an Ull (3).** NM 6974 7178. Named *Meall an Aoil* 'the hill of the lime', with gen. sg. of *aol* m. 'lime'. 'limy hill' on OS maps.

Eilean Dubh an t-Sailean Beag and Eilean Dubh an t-Sailean Mòr from Port a' Bhata

8. **I Dou Tallan Beg (5).** NM 6917 7245. G. *Eilean Dubh an t-Sailean Beag* 'the small black island of the inlet'.

9. **I Dou Tallan more (5).** NM 6924 7238. G. *Eilean Dubh an t-Sailean Mòr* 'the large black island of the inlet'. Named *Eilean Dubh* 'the black island' on OS maps. Both islands may have formed part of *Port a' Bhata*, especially if the crossing was easier from that side.

10. *Ru Maul an Ull (3).* NM 6980 7218. G. *Rubha Meall an Aoil* 'the promontory of *Meall an Aoil'*. See no. 8, above.

11. *S. an Lochan (3).* NM 7000 7215. G. *Sàilean an Lochain* 'the bay of the lochan', with gen. sg. of *lochan* m., diminutive of *loch* m. 'loch'.

12. *Loch an c. dhuin (3).* NM 7005 7130. Named *Lochan Fheith Dhuinn* on OS maps.

13. *Ru Innes Ru (5).* NM 6980 7214. Potentially G. *Rubha Innis an Rubha* 'the promontory of *Innis an Rubha* (the island of the promontory)'.

14. *Rubh Tòrr na Slingrich (5).* NM 716 695. G. *Rubha Tòrr na Slinneanach* 'the promontory of the knoll of the broad-shouldered man' or *Rubha Tòrr na Sligearnach* 'the point of the knoll abounding in shells', although it is an unlikely place to find shells. Named *Rubha Tòrr na Cachlaidh* on OS maps.

15. *Tom an Aoil (3).* NM 7165 6983. G. *Tom na Aoil* 'the hill of the lime', with gen. sg. of *aoil* m. 'lime'. Named *Tòrr an Slingrich* on OS Maps.

16. *Croit Iain Iseobail (64).* NM 7025 6958. G. *Croit Iain Iseobail* 'the croft of *Iain* and *Iseobail* (MacEachen)', who lived there between 1870 and 1920. The house has now been reconstructed and named '**Toad Hall**'.

17. *Taigh an Fhigheadair (64).* NM 7035 6974. G. *Taigh an Fhigheadair* 'the house of the knitter'. This is now 'The Moidart Smokehouse', and was once occupied by and named after a lady knitter called Peggy MacDonald.

Mullen Ull beside the *Allt a' Mhuillin*   © Dr. Sandra Evans

18. **Mullen Ull (49). NM 6929 7198.** When Alexander Macdanald of Glenalladale bought Langal in 1813, the Feu Charter mentions *The whole town and lands of Delelea, Langall <u>excepting Mullen Ull</u>*...etc. It is assumed that this name refers to the mill on the *Allt a' Mhuillin,* which was constructed after 1760, and the name suggests that it was built using lime mortar, with *aoil* 'lime'. The mill is actually on Port a' Bhata land, although traditionally it appears to have been on Langal. A similar situation occurs in Sunart, where the mill was 'always' on Ranachanmore, but in it's final reconstruction, was built across the burn on Ranachanstrone (but see 19 below).

A developed shieling, with lazybeds enclosed by a turf and stone dyke

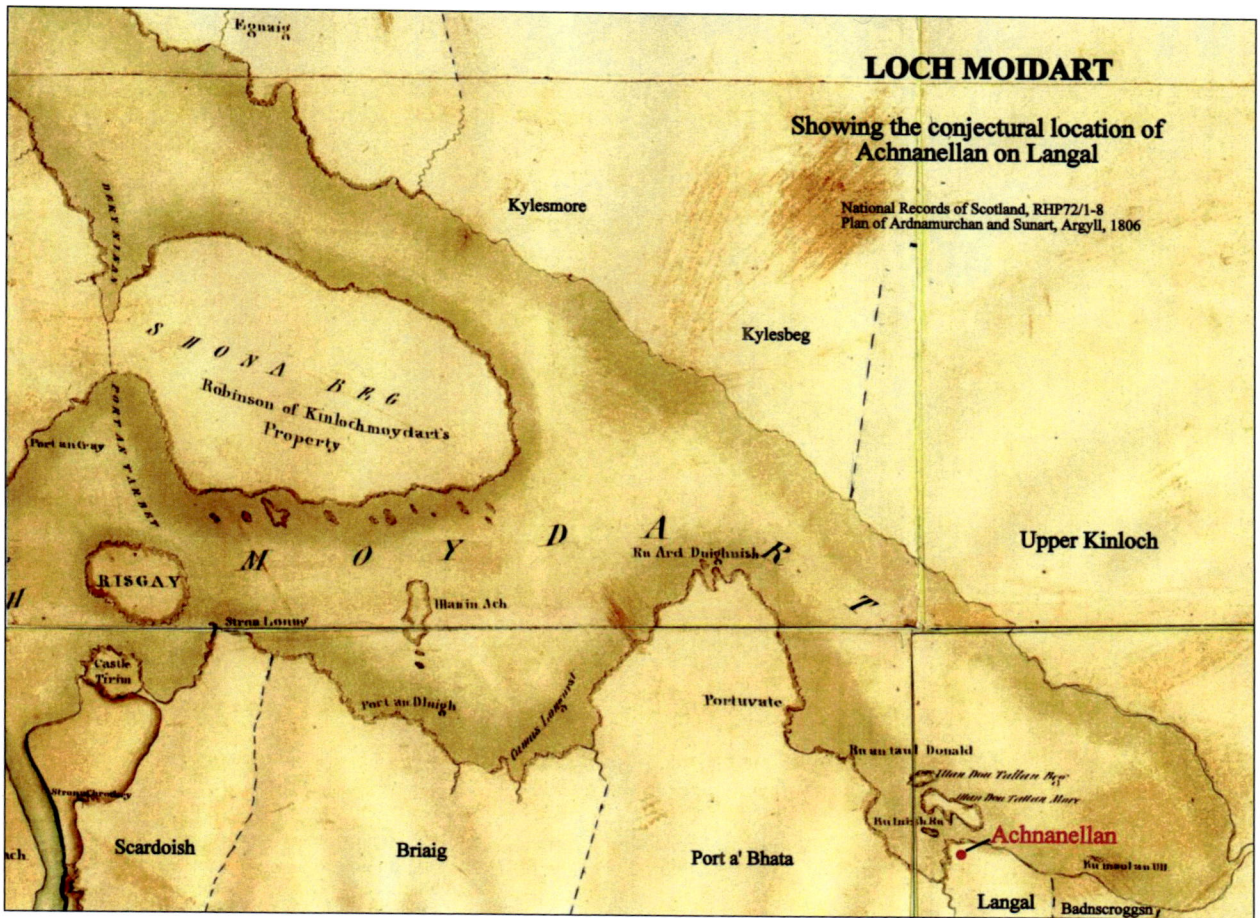

LOCH MOIDART

Showing the conjectural location of Achnanellan on Langal

National Records of Scotland, RHP72/1-8
Plan of Ardnamurchan and Sunart, Argyll, 1806

19. **Auchnellan (30). NM 6930 7270 (?).** G. *Achadh nan Eilean* 'the field of the islands'. The islands were probably *Eileanan Dubh an t-Sailean* in Loch Moidart. Auchnellan was described as a one merkland in 1625, and equating it with this location is purely conjectural, but it was almost certainly an alternative name for Langal township at this time, and not Achanellan in Sunart. There are at least three corn kilns within the settlement, suggesting that this was the croft attached to the corn mill 'Mullen Ull'. Alternatively, Mullen Ull could be derived from *Meall an Aoil* and used as a later name for the Mill Croft or 'Auchnallan'. There may be two tidal fish traps associated with the islands.

20. **Bad Air Rais (19). NM 719 702.** G. *Bad Air Rais* ' the thicket on the path'.

21. **Pit an Clachan Dearag (75). NM 7168 7011.** G. *Pit an Clachan Dearg* 'The pit of the red stone'. The late Roddie (Langal) MacDonald said that this was the name of a borrow pit beside the main road near The Old Poor's House.

22. **Pairce Dubh (75). NM 7069 6952.** G. *A' Phàirc Dhubh* 'the black park'. Probably a field taken in from the moss.

DRUM LOY ; Druim an Laoigh : the ridge of the calf

1625, Drumnaleiwe; Camistrollane (30); 1700, Driminle (58); 1718, Drumnaloy (61); 1739, Drimnlaoigh (1); 1744, Drimluy (38); 1748, Dreimluoy (32); 1760, Driminloy, Drimnloy (42); 1782, Drimlaoi (64), Drimintui (65); 1786, Drimlu (56); 1790, Easter Drum<u>ly</u>, Wester Drum<u>by</u> (48); 1823, Drumloy (7).

Wester Drumloy   Easter Drumloy

Valuation: A ½ merk land in 1625 (plus the ½ merk land of Camistrollane (?), a 1 merk land in 1718, 1748 and 1760. 'Easter Drum<u>ly</u>' and 'Wester Drum<u>by</u>' were both half merk lands in 1790.

Area: C. 1816 there was 3.99 acres of agricultural land in 3 fields and 1153.20 of moorland, total 1157.19 acres. The field at *An Garbh Allt* and the developed shieling on the *Allt Beag* appear to post-date this measurement.

Tenants: 1625, wadset to Johne ronnaldsoun persoun of Ellanfinnan for 19 years, then to his 'brother soun' Allane McRonald (30); 1691, Duncan McVarrish and his spouse Christian NcDonald and their children Eune, John and Donald, (58); Wadset to Maighstir Alasdair McDonald who died in 1718, then to Aneas McDonald of Dallellie till after 1748. John McVarish, chirurgeon in Moidart in 1740 (12); 1744, Donald McVarish and Cathrine McDonald his spouse, John McDonald, Euan McDonald and Duncan McDonald, all tennants in Drimnloy, and John mc Innes of the ferry of Island Finnan equally between them all (42). John McVarish got the tack of one Farthing land for 12 years from 30 March, 1786 (56); 1782, John MacVarish (64). 1790, in tack to Alexander Macdonald of Dalelia (48). 1813, Drumloy was sold to Alexander Macdonald of Glenalladale (49).

Settlements: NM 7685 6988. *Druim an Laoigh.*
NM 7880 6988. *An Garbh Allt.*
NM 7821 7094. 'Fafsy Ganoir'.

Shielings: NM 7703 7141. *Allt Beag Druim an Laoigh.* Developed shieling with house and enclosed field system.

Placenames:
1. **Drumloy (5. NM 7685 6988.** G. *Druim an Laoigh* 'the ridge of the calf', with gen. sg of *laogh* m. 'calf'.

2. **Ru Ard Monal (5). NM 7656 6938.** G. *Rubha Àird a' Mhuineil* 'the promontory of the headland of the neck (of land)', with gen. sg. of *muineal* m. 'neck'. Named *Rubha na h-Àirde* 'the promontory of the headland' on OS maps.

3. **Ru Ghannach (5). NM 7707 6926.** Perhaps for G. *An Rubha Gainmhich* 'the sandy promontory'.

4. **Achie in Unie (5, 36). NM 7774 6926.** G. *Achadh an Aonaich* 'the field of the hill', with gen. sg. of *aonach* m. 'hill, moor, meeting place'.

5. **Torr a Bann (34). NM 7714 7000.** G. *An Tòrr Bàn* 'the fair hill'.

6. **Cruoch na Gaul (3). NM 7572 7193.** G. *Cruach nan Gall* 'the hill of the strangers', with gen. pl. of *gall* m. 'stranger, foreigner, Lowlander'.

7. **Cruach na Tolach (3). Cruach na Totach (34). NM 7615 7226.** G. *Cruach na Tollaiche* 'the hill of the holed ground', with gen. sg. of *tollach* f. 'holed ground, place of holes; full of holes'.

8. Cruoch Gorme (34). NM 7677 7224. G. *Cruach Gorm* 'the blue (green) prominent hill'.

One of the buildings on Fafsy Ganoir, with the Bronze Age burial cairn forming the knoll

Fafsy Ganoir from the burial cairn, looking to the NE

9. Fafsy Ganoir (3). NM 7821 7094. G. *Fasadh G...?* 'The name of a dwelling with an unknown specific, but possibly a local word for a giant. A rich field enclosed by a boulder dyke, and with deeply cut lazy-beds. Within is a large Bronze Age burial cairn, at least four buildings, possibly the largest corn-kiln in the district and a charcoal-burner's platform overlying lazy-beds. Named *Fàsadh an Fhamhair* 'the dwelling of the giant' on OS maps. Was this how local people saw the cairn at one time; the abode of a giant?

10. Ru Fassyganoir (5, 36). NM 7822 7093. G. *Rubha Fàsadh* Cf. no. 5.

11. **Ru Tom an Tleachk (36).** NM 7586 6227. *Rubha Tom an t-Sluichd* 'the headland of the hill of the hollow or dell'.

12. **Camistrollane (30).** Perhaps G. *Camas an Drollamain* 'the bay of the labourer', with gen. sg. of *drollaman, dreallaman* m. 'unskilled labourer'. The Tack of Dalelia and other lands granted to the Parson of Island Finnan in 1625 included the ½ merk land of Camistrollane which presumably takes its name from *Camas Drollaman*, **NM 7640 6950.** The settlement of 'Camistrollane' probably consisted of the buildings and land to the west of the *Allt Druim an Laoigh* at **NM 7621 6989.** Latterly **Wester Drumloy.**

13. **Craig an lach (34).** NM 7615 7150 (?) Possibly a mistake for *Creagan an Fhithich* ' the small crag of the raven' on Dalelia.

14. **Easter Drumly (48).** NM 7670 6989. Easter Drumloy or 'Drumnaleiwe' in 1790.

15. **Wester Drumby (48).** NM 7680 6990. Wester Drumloy or 'Camistrollane' in 1790.

The ridge of *Druim an Laoigh* after which the township was named

Loch Shiel from Fafsy Ganoir

250

LANDS IN ARISAIG BELONGING TO MACDONALD OF BELFINLAY, 1739.

THE BELFINLAY ESTATE

Angus McDonald, 2nd of Belfinlay held a small estate on Benbecula, but exchanged it for one in Arisaig in 1720. The Arisaig estate comprised Pendui, Laggan, Essan, Allasary, Torbay, Ranachan, Moy and Peinmeanach. Angus was succeeded by his elder son Donald, who died without issue and followed by his brother Ranald, 4th of Belfinlay, a Captain in the Clanranald Regiment during the '45. Ranald died in 1749 and was succeeded by his uncle Allan, 5th of Belfinlay. The following townships formed part of the 24 merk lands of Arisaig, although they appear to be geographically in Moidart.

INVERAILORT : Inbhir Ailleart : the mouth of (the river of) *'Ailleart'* or KINCREAGAN : Cinn a' Chreagain : the end of the rock

1713, Arian (6); 1762, Kenichregen (53); 1775, Kinchregan (47); 1782, Kennachreggan (64); 1786, Kenchrecan, Kenichrecan (55); 1798, Kenachreggan (10); 1823, Kenchreggan (7); 1832, Kenachreggun (34); 2002, Inverailort (2).

Arian Fank, almost certainly on the site of the settlement

Valuation: Arian was a ¾ pennyland in 1718. Kincreagan was a three farthing land in 1775.

Tenants: 1718, John McColl Aird held a 2 farthing land of Arien, with a 1 farthing land lying waste (6); 1762, set in tack to Allan McDonald of Bellfindlay for 13 years (52); 1782, Ewen MacDonald (64); 1777, Colin MacDonald of Boisdale received a tack for 11 years (47), which was renewed for a further 19 years in 1786 (55). 1798, Kincreagan, Arian and Essan were let to Angus Macdonald, sub-tenant of Colin Macdonald of Boisdale (71).

Settlements: NM 7804 8195. Arien.
                     NM 7646 8150.

Shielings: Not recorded.

Placenames:
   1.   **Inverailort (2). NM 7692 8225.** G. *Inbhir Ailleart* 'the mouth of (the river of) *Ailleart'.*

Burial ground of the Cameron-Head family

2. **Kenacreggan, (10). NM 7652 8157.** G. C*inn a' Chreagain* 'the end of the rock', with gen. sg. of *creagan* m. 'rock, hillock'.

3. **Allt Raostig (17). NM 7750 8216.** Perhaps G. *Allt Raostaig* 'the burn of R..', with an obscure Old Norse loan-name. It is unlikely to be G. *Allt Roistich* 'the roach burn'. Named 'Roti Burn' on OS maps.

4. **Ault Gorme (34). NM 7718 7800.** G. *Allt Gorm* 'the blue (or green) burn'. Named *Allt a' Bhuiridh* on OS maps.

5. **Arien (34). Ariane (52). NM 7804 8195.** G. *Àiridh Shithean* 'the shieling of *Gleann Sithean* (the fairy hill)', *Àiridhean* 'the shielings' or *Àiridh Iain* 'Iain's shieling'. The halfpenny land of Ariyen was set in tack to Colin Macdonald of Boisdale in 1775, for a period of 11 years (47). Boisdale's tack was extended by a further 19 years in 1786, when Arien was valued at four farthings (55).

6. **Backnicharey (46).** Location not known. G. *Bac na Caraidh* 'the bank of the fish trap'. "The halfpenny land of Ariyen, the halfpenny land of Essan and the three farthings land of Kinchregan with the two penticles of Upper and Lower Backnichareys if not already in tack (set to Colin Macdonald of Boisdale) for 11 full years with right to manufacture kelp on all the shores and half the fishing on the River Aillort'(47). When Boisdale's tack was extended in 1786, it included 'the grassing of Burie and Bachnahaiy or Bachnahary' (55). Burie could be Blarbuie, while Bacnahary, *Bac na h-Àiridh* 'the bank of the shieling' may be a better interpretation than *Bac na Caraidh.*

7. **Blarbuie (53.** G. *Blar Buidhe* 'the yellow field'. Location not known. There was a tack to Allan McDonald of Bellfindlay dated May 11th, 1762, for the halfpenny land of Blarbuie and the farden (farthing) land of Kinichregan with the mansion house thereon for the space of 13 years from Whitsunday next. By 1775, Blarbuie and Kinichregan were combined to form the three farthings land of Kinchregan. Could *Blar Buidhe* have been an alternative name for Easter Alisary?

8. **Maol Taff (34). NM 7923 7977 (?)** G. *Meall Damh* 'the bare round hill of the stag'. This may have been an earlier name for *Beinn Coire nan Gall* ' the mountain above the corrie of the strangers'. *Meall Damh* is shown as the shoulder of the hill at NM 7865 7990.

9. **Gleann Sithean (19). NM 7717 8000.** G. *Gleann Sìthean* 'the glen of the fairy hill'.

10. **Glen Ailort (19) . NM 7800 8300.** G. *Gleann Aileort.* This name is recorded in early Registrar Records.

Arian fank looking NW

Innis na Cuilce burial ground, looking NE and grave-slabs covered in bracken

11. Innis na Cuilce (2). Eilean na Cuilc (19). NM 7725 8237. G. *Innis na Cuilc; Eilean na Cuilc* 'the island of reeds', although not a suitable habitat for the reed *Phragmites communis.* A very small island in the River Ailort, on the march between Inverailort and Kinlochailort, and subject to periodic innundation. The highest part is slightly elevated above the level of spates, enclosed by a dyke and for many years used as a burial ground, mainly for the people of Ardnish. The most recent interrment appears to have been in the 1970s. Most of the graveslabs are recumbent, and covered with bracken.

Pools on the River Ailort, 1908.

Lower Beat.
12. **New Bridge Pool (9). NM 7650 8199.** Also known as **Sea Pool**.
13. **Butts Pool (9). NM 7665 8193.**
14. **Mrs Cameron-Head's Pool (9). NM 7674 8212.**
15. **Bridge Pool (9). NM 7687 8232.**
16. **Policeman's Pool (9). NM 7716 8225.**
17. **Rock Pool (9). NM 7734 8268.**
18. **Lord Elgin's Pool (9). NM 7761 8291.**
19. **Monument Pool (9). NM 7786 8296.**

Upper Beat.
20. **Deep Pool (9). NM 7791 8302.**
21. **Round Pool (9). NM 7797 8303.**
22. **McPherson's Pool (9). NM 7805 8303.**
23. **South Bend Pool (9). NM 7807 8305.**
24. **Run Out (9). NM 7855 8340.**
25. **Srean (9). NM 7855 8340.** Perhaps contains either G. *sruthan* 'burn' or *srian* 'stripe'.
26. **Falls (9). NM 7879 8290.**
27. **Sluice Run (9). NM 7886 8275.**
28. **Castle Island (9). NM 7973 8248.** Named ***Eilean an Taighe*** 'the island of the house' on OS maps, with gen. sg. of *taigh* m. 'house'.
29. **Duncan's Burn (9). NM 7675 8195.**

Upper Loch Ailort looking SW from Lower Polnish to Rois-bheinn and the rough grazing on Easter Alisary

ALISARY Aill Assairidh

1748, Ifiroy (28); 1782, Alassary, Alasary (64); 1823, Allisary (7); 1827, Alisary (50); 1832, Allisarry (34); 2002, Alisary (2).

West Alisary fank

Valuation: A ½ merk land and a 6/8d land in 1748 and a ½ penny land in 1823.

Tenants: Donald McDonald (Domhnuill Ailisary), 3/4d, Hugh Smith, 1/8d and Hugh McDonald, 1/8d in 1748 = 6/8d for the ½ merk land; 1772, Donald MacDonald Alisary emigrated to Prince Edward Island. 1782, John MacInnes, Alexander Maclean (64). In 1798 it was let to Angus Macdonald and Alexander Maclean at a rent of £70-0-0 (71). Tack between Clanranald and Trustees to Capt. Ranald McDonald for 13 years from 1821 (39). Feu disposition in favour of Major Allan Nicolson Mac donald, 1827 (50).

East and West Alisary from above the Irine Burn

Settlements: NM 7428 7954. Wester Alisary. At least four buildings.
NM 7443 7963. Easter Alisary. At least four buildings.

Shielings: Not known.

Placenames:

1. **Alisary (2). NM 7431 7964.** G. *Aill Assaraidh ?* When the ''Roshven Estate' was sold in 1827, it included 'all and whole the western part of Alisary being part and portion of all and hail the 24 merk land of Arisaig' (50).

A *port* on West Alisary

2. **Allt a' Phìobaire (18). NM 7537 8100.** G. *Allt a' Phìobaire* 'the burn of the piper', with gen. sg. of *pìobaire* m. 'piper'. Named *Allt Sròn an Uinnseinn* 'the burn of the point of the ash tree' on OS maps. *Pathfinder 262*

3. **Sròn an Uinnseinn (13). NM 7515 8147.** G. *Sròn an Uinnseinn* 'the point of the ash tree'.

4. **Allt Rutha an Uinnsinn (19). NM 7537 8100.** G. *Allt Rutha an Uinnsinn* 'the burn of the ash tree headland'. *Rutha* appears to be an earlier or alternative form of *Rubha*. *also Rudha (Norse origin. Angus Henderson.)*

5. **Bealach Breac (21). NM 7465 8048.** G. *Am Bealach Breac* 'the speckled pass'. This was the highest part of the old road between Kinlochailort and Alisary. The Ordnance Survey place *Bealach Breac* at NM 7521 8027.

6. **Sgur Corrie rioch (34). NM 7505 8037.** G. *Sgùrr a' Choire Riabhaich* 'the mountain of the tawny corry', with gen. sg. of *coire* m. 'corry'.

An Stac on East Alisary

ESSAN : An t-Essan : the waterfall
1625, Essan (74); 1775, Essan (47); 1823, Issan (7); 2002, Essan (2).

Essan, at the foot of the Allt Easain

Valuation: A ½ penny land in 1776 and <u>tow</u> farthings in 1786.

Tenants: Colin Macdonald of Boisdale was tacksman in 1775 for 11 years (47) and in 1786 for a further 19 years (55). In 1798 it was sub-let to Angus Macdonald (71).

Settlements: NM 8179 8173. *Essan.* An improved shepherd's cottage with at least two earlier houses in close proximity.

Shielings: Not recorded.

Placenames:

1. **Essan (2). NM 8179 8173.** G. *An t-Easan* 'the waterfall', with *easan* m., diminutive of *eas.* m. 'waterfall'. In 1625, Ranald MacDonald, 2nd son of Allan, 8th of Clanranald, received from his father a number of properties in Benbecula, together with Ardness, Lochylt and Essan in Arasaig (74).

2. **Coire an t-Sassanaich (19). NM 805 810.** G. *Coire an t-Sasanaich* 'the corrie of the Englishman', with gen. sg. of *Sasanach* m. 'Englishman'. A slab in the corrie marks the grave of a Redcoat shot here by a Moidart man in 1746.

3. **Ault na larugh (34). NM 8234 8100.** G. *Allt na Laraig* 'the burn of the ruin'. Named *Allt na Diollaid* 'the burn of the saddle' on OS maps. *Diollaid* may however be derived from *dubh* 'black, dark' and *leitir* 'slope', giving G. *Allt an Dubh-Leitir* 'the burn of the dark slope'. i.e. a cold north-facing slope with little or no sun in winter.

    Shape of the hill.

4. **Diled facnach (34). NM 8179 8087.** This is shown on the hill overlooking Essan, but the location is uncertain. 'Diled' is almost certainly G. *Diollaid* or *An Dubh-Leitir* but 'facnach' is obscure.

5. **Diallod (19). NM 8179 8173.** G. *Diollaid* 'the saddle'. This name was sometimes used as an alternative to *Essan,* with *Diollaid Mòr and Diollaid Beg*; was this latter name also used for *A 'Mhuic* below?

257

The *Allt na Diollaid* and the waterfalls that provided the township name of *An t-Easan*

MUIC : A' Mhuic : the rick

This name is written in pencil on the 1:10,560 scale 1st Edition OS map depicting the Inverailort Estate. The boundaries are uncertain and it may never have functioned as a township. Perhaps it was a shepherd's croft on Essan in the latter part of the 19th century.

Settlement: At least two buildings to the west of the *Allt a' Bhuide Choire* and a fank on the eastern side.

Choire Bhuidte

Placenames:

1. **Muic (20). NM 8345 8166.** G. *A' Mhuic,* dative of *A' Mhuc* 'the rick or haystack'. It could just as easily be *Muic,* the same name as a township on Loch Arkaigside, usually translated as 'pig place'.

2. **Ballach lean (34). NM 8509 7845 (?)** G. *Bealach Leathan* 'the broad pass'. Access to the ridge is difficult on both sides and it is an unlikely place for a pass.

3. **Craigarrie ghour (34). NM 8516 8125.** G. *Creag an Àiridh Ghobhar* 'the rock of the goat shieling'. *Creag Ghobhar* 'the goat rock' is shown on OS maps.

UNLOCATED NAME.

1. **Powait (7).** The town and land of *Powait* was listed along with other townships in 1823. Possibly a mis-reading of G. *Diollaid.*

258

REFERENCES

1. NAS, TE19/16 4a3. Rental of the Parish of Ardnamurchan, 1739.
2. OS Explorer 390. *Ardnamurchan, Moidart, Sunart and Loch Shiel;* 1:25,000 scale map (2007).
3. Plan of part of Moidart the property of Alexr. MacDonald Esq. Lochshiel Estate Plan of c. 1816.
4. Admiralty Chart No. 531, Loch Moidart, 1860, surveyed by James Jeffery Master R.N.
5. NLS Ms 384 (1). Moidart names on RHP 72.
6. Rixson, Denis (Editor). 2008. *Rentals for Moidart and Arisaig, 1718.* Mallaig Heritage Centre, based on GD 201/5/1257/1 and 2.
7. Excerpta e Sasinarum Registris Vice-Comitatum de Inverness Vol. II no.172 (1823).
8. Excerpta e Sasinarum Registris Vice-comitatum de Inverness Vol. IV, no 386 (1845).
9. Dressler, C. and Stiùbhart, D. W. (2012). *Alexander MacDonald, Bard of the Gaelic Enlightenment.* The Islands Book Trust.
10. NRS, GD128/49/3/1
11. MacDonald, Rev.A and MacDonald, Rev.A (1904). *The Clan Donald* Vol.III p.655. Contract between Donald MacDonald of Clanranald and Roderick MacDonald of Glenaladale, 1674.
12. McDonald, David (2003). *Clanranald; A History of the Clanranald Regiment 1745-46,* from contemporary sources. Bedford.
13. Ordnance Survey. Explorer Map, Sheet 398, *Loch Morar & Mallaig;* 1:25,000 scale, (2009).
14. Clanranald Estate Sales Plan,
15. Oral Tradition of the District.
16. Ordnance Survey 1:10,560 Map, 2nd Edition.
17. Ordnance Survey 1:10,560 Map, 1st Edition, Inverness-shire, Sheets CXLVII and CXLVIII, surveyed 1873 by Capt Coddington, RN.
18. Highland Council Roadsign but also in 14.
19. Oral Tradition; Tearlach MacFarlane, 2009, 2014 .
20. Inverailort Estate Plan on 1:10,560 scale OS map. Manuscript additions to Lochaber Archives, Fort William.
21. Leigh, Margaret (1950). *Spade Among the Rushes.* Readers Union with Phoenix Press.
22. MacDonald, Father Charles (1889). *Moidart,* or *Amongst the Clanranalds.* 1997 reprint edited by John Watts. Birlinn.
23. Wood, Wendy (1950). *Moidart and Morar.* The Moray Press.
24. Oral Tradition; Gordon Barr, 2010.
25. Oral Tradition, Alasdair (Pod) Carmichael, 2010.
26. Oral Tradition, Angus Peter Maclean, 2010.
27. Oral Tradition, Jeff Carr, 2010.
28. NRS, E764/1/2. Judicial Rental of the Estate of Kenlochmoidart by David Bruce, 1749.
29. NRS, GD241 Box 170 Bundle 6, No. 7.
30. MacDonald, Rev. A and Rev.A (1900). *The Clan Donald,* Vol. II, pp.773-7. Tack by John MacDonald of Clanronald to the Person of Island Finnan and others, 1625.
31. William Roy's Military Survey of Scotland, 1747-55. British Museum.
32. E744/1/2. Judicial Rental of the Estate of Clanranald by David Bruce, 1749.
33. Wood, Wendy (1946). *Mac's Croft.* Fredrick Muller Ltd, London.
34. John Thomson's 'Atlas of Scotland, 1832'. Map of Inverness-shire. National Library of Scotland.
35. Excertae Sanisarum Registris Vice-comitatum de Inverness, Vol. , No. 943, (1829) and Vol. V, No. 436 (1850).
36. NRS, RHP 72. William Bald's Survey of Ardnamurchan and Sunart, 1806-7.
37. Ordnance Survey, Explorer Map 391, Ardgour and Strontian, 1:25,000 scale (2002).
38. Presbytry of Mull Records, 1744. Argyll CC Archives. Minutes relating to Mr. Francis MacDonald's alleged incestuous relationship with his sister.
39. NRS. GD201/5/1196. Tack of townships in Eilean Shona, 1788.
40. NRS. GD201/5/1204. Tack of Irene and West Allisary, 1821.
41. NRS. GD201/5/1190. Tack of Samalaman, 1786.
42. NRS. GD201/2/23. Tack of Drumloy, 1760.
43. NRS. GD201/2/24. Tack of townships in Eilean Shona, 1760.
44. NRS. GD201/2/26. Tack of Briaig and Blain, 1760.
45. NRS. GD201/2/27. Tack of Scardoish, 1760.
46. NRS. GD201/2/47. Tack of Lochans, 1773.
47. NRS. GD201/2/49. Tack of Arian, Essan and Kincreagan, with the 2 penticles of Upper and Lower Backnachareys, 1775.
48. NRS. GD201/2/58. Tack of Dalelia, Langal, Easter and Wester Drumloy and Annat, with the Island Finan Ferry and Changehouse, 1790.
49. NRS. GD201/2/65. Feu charter of Dalelia, Langal, Annet, Drumloy, Mingary, Blain, Eilean Shona, Breaig and Portavata, 1813.
50. NRS. GD201/2/68. Feu charter of Irene, Glenuig, Samalaman, Smerisary, Egnaig and Western Alisary, 1827.
51. NRS. GD201/5/1152. Tack of Mingary, 1760.
52. NRS. GD201/5/1154. Tack of Ariane, 1762.
53. NRS. GD201/5/1156. Tack of Kincreagan and Blarbuie, 1762.

54. NRS. GD201/5/1159. Tack of Inchrory, 1762.
55. NRS. GD201/5/1189. Tack of Kencreagan, Arien and Essan, with the grassings of Burie and Bachnahary, 1786.
56. NRS. GD201/5/1188. Tack of Drumloy, 1786.
57. Grant, FrancisJ. Editor ((1902). *The Commissariot Record of Argyle; Register of Testaments, 1674-1800.* Scottish Record Society, Edinburgh..
58. Grant, Francis J. Editor (1909). *The Commissariot of Argyll; Register of Inventories, 1693-1702.* Scottish Record Society. James Skinner and Co, Edinburgh. 1909.
59. Dye, John (Editor), 2000. *Some Priests of Moidart* by Father Jerome Ireland; (first printed in the weekly Moidart Parish Newsletter, 1968-70). Blackfriars Publications, Chapel-en-le Frith, Derbyshire.
60. OS 25 inches to the mile map of Loch Shiel. Forestry Commission Survey of the proposed road along the side of Loch Shiel.
61. NRS. GD201/5/1257/1. Judicial Rental of the Real Estate of Ranald McDonald, Late Captain of Clanranald, 1718.
62. NRS. GD201/6/6.
63. Cregeen, Eric (1964) *Argyll Estate Instructions, Mull, Morvern, Tiree, 1771-1805,* (ed.). Scottish Records Society.
64. NLS Ms. 3983 f. 44. Robertson MacDonald of Kinlochmoidart Collection. Lt. Col. Alexr. MacDonald of Kinloch; Roup dated 20/5/1782.
65. NLS Ms. 3983 f. 63. Robertson MacDonald Papers. List of persons sent bills for above.
66. NLS Ms. 3983. f. 20. Robertson MacDonald Papers. Grazings on Craig, Badnachraggan and Tengaglas.
67. NLS Ms. 3893. Robertson MacDonald Papers. People that collected kelp on the estate.
68. NLS Ms. 3893. Robertson MacDonald Papers. Rentals of Kinlochmoidart, 1786.
69. NLS Ms. 3983. Robertson MacDonald Papers. F. 81, 83, 88, 90, 91. Tacks granted to various people.
70. NLS Ms. 3983. Robertson MacDonald Papers. Ms. f. 171. List of tenants in 1802.
71. Judicial Rental of Clanranald, 1798. GD 201 /5/216.
72. NLS Ms. 3893. Robertson MacDonald Papers. Rentals of the Estate of Kinlochmoidart, 1755.
73. King, J. and Clyne, H. (2013). *The Rough Bounds of Lochaber.* Scottish Natural Heritage/Ainmean-Àite na h-Alba.
74. MacKenzie, Alexander (1881). *The MacDonalds of Clanranald.* Inverness; A & W. MacKenzie.
75. Oral Tradition; John Dye, 2014.
76. Imrie, John (Ed) 1969. *Judiciary Records of Argyll and the Isles, 1664-1742.* Stair Society, Vol. 25. pp. 404-424.
77. Census Returns, 1881.
78. Oral Tradition: Iain MacMaster, Mingary.
79. Dressler, C. and Stiùbhart, D.W. (2012), *Alexander MacDonald, Bard of the Gaelic Enlightenment* The Islands Book Trust.
80. *Register of the Privy Council*, First Series Vol. XI p. 149.

The cairn beside the *Garbh Allt,* Annat

INDEX OF MOIDART PLACE-NAMES AS RECEIVED
TOWNSHIP NAMES IN BOLD

263

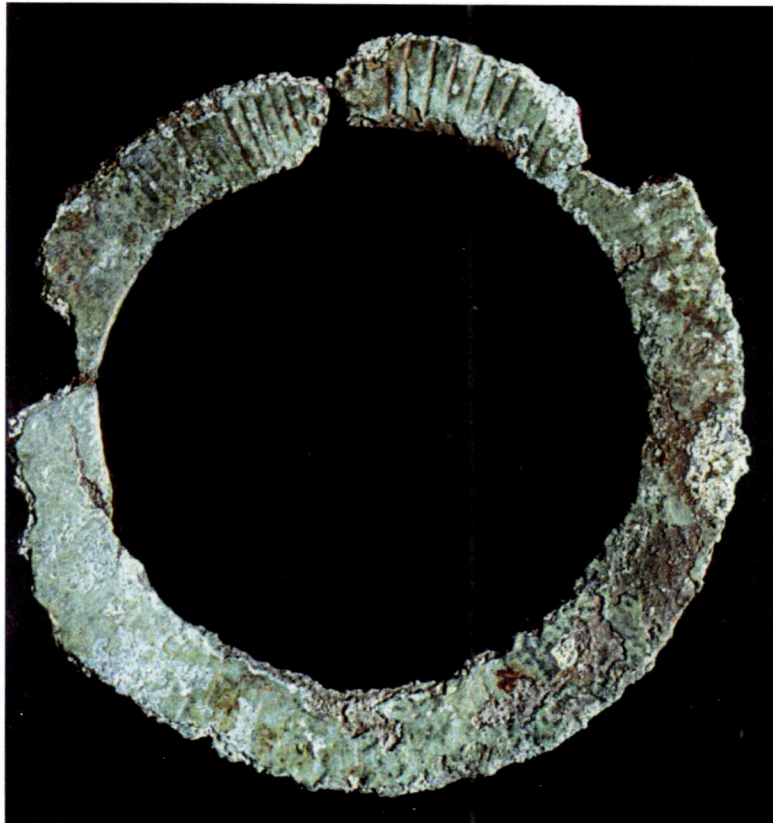

Copper alloy brooch found on *Eilean Fhìonain* 1895/6 © National Museums Scotland

INDEX OF GAELIC NAMES IN MOIDART

Asterisk denotes hypothetical forms
Township names in bold

BIBLIOGRAPHY

Black, Ronald (1986). *Mac Mhaighstir Alasdair; The Ardnamurchan Years.* The Society of West Highland and Island Historical Research.

Black, Ronald Editor (2001). *An Lasair: Anthology of 18th century Scottish Gaelic Verse.* Birlinn.

Campbell, J.L Editor (1958). *Gaelic Words and Expressions from South Uist and Eriskay* collected by Rev. Fr. Allan MacDonald of Eriskay (1859-1905). Dublin Institute for Advanced Studies.

Cregeen, Eric (1964). *Argyll Estate Insructions: Mull, Morvern, Tiree, 1771-1805, (Ed),* Scottish Records Society. T. and A. Constable Ltd.

Dressler, C. and Stiùbhart, D. W. (2012). *Alexander MacDonald, Bard of the Gaelic Enlightenment.* The Islands Book Trust.

Dwelly, Edward (1901-1911). *The Illustrated Gaelic-English Dictionary.* (Ninth Edition, 1977). Gairm Publications, Glasgow.

Dye, John, Editor (2000). *Some Priests of Moidart* by Father Jerome Ireland; first printed in the weekly *Moidart Parish Newsletter, 1968-70).* Blackfriars Publications, Chapel en le Frith, Derbyshire

Ferguson, Robert (1856). *The Northmen in Cumberland and Westmoreland.* London: Longman and Co. and Carlisle: R. & J. Steel.

Grant, Francis J. Editor (1902). *The Commissariot Record of Argyle; Register of Testaments, 1674-1800.* Scottish Records Society, Edinburgh.

Grant, Francis J. Editor (1909. *The Commissariot of Argyll; Register of Inventories, 1693-1702.* Scottish Records Society. James Skinner and Co., Edinburgh.

King, J. and Clyne, H. (2013). *The Rough Bounds of Lochaber.* Scottish Natural Heritage/Ainmean Àite na h-Alba.

Leigh, Margaret (1950). *Spade Among the Rushes.* Readers Union with Phoenix Press.

MacAlpine, Neil and MacKenzie, John (1979). *Gaelic/English and English/Gaelic Dictionary.* Gairm Publications.

MacDonald, Rev. A and Rev. A (1900, 1904). *The Clan Donald, Vols II and III.*

MacDonald, David (2003). *Clanranald; A History of the Clanranald Regiment, 1745-46,* from contemporary sources. Bedford.

MacDonald, Father Charles (1889). *Moidart; or Amongst the Clanranalds.* 1997 reprint edited by John Watts. Birlinn Press.

MacDonald, Norman H. (2008). *The Clan Ranald of Garmoran. A History of the MacDonalds of Clanranald.* Edinburgh.

MacKenzie, Alexander (1881). *The MacDonalds of Clanranald.* Inverness: A. & W. MacKenzie.

Maclean, Alasdair (2001). *Night Falls on Ardnamurchan; the Twilight of a Crofting Family.* Birlinn.

Maclean-Bristol, N. Editor. (1998). *The Inhabitants of the Inner Isles, Morvern and Ardnamurchan, 1716.* Scottish Records Society. New Series Vol. 21. Edinburgh.

Rixson, D. (Editor). 2008. *Rentals for Moidart and Arisaig, 1718.* Mallaig Heritage Centre, based on GD 201/5/1257/1 and 2.

Todd, J. H. *Martyrology of Donegal.* Irish Archaeology Society.

Thomas, Charles (1971). *The Early Christian Archaeology of North Britain.* Oxford University Press.

Watson, W. J. (1926). *The Celtic Place-names of Scotland.* Edinburgh and London. 2004 Edition, Birlinn Ltd.

Wood, Wendy (1946). *Mac's Croft.* Fredrick Muller Ltd, London.

Wood, Wendy (1950). *Moidart and Morar.* The Moray Press.

Architectural fragments from the old parish church of *Eilean Fhìonain*